Parametric Estimating for Executives and Estimators

Parametric Estimating for Executives and Estimators

Paul F. Gallagher
Estimating and Cost Analysis Consultant

VNR VAN NOSTRAND REINHOLD COMPANY
NEW YORK   CINCINNATI   TORONTO   LONDON   MELBOURNE

Van Nostrand Reinhold Company Regional Offices:
New York   Cincinnati

Van Nostrand Reinhold Company International Offices:
London   Toronto   Melbourne

Library of Congress Catalog Card Number: 81-1569
ISBN: 0-442-23997-1

Manufactured in the United States of America

Published by Van Nostrand Reinhold Company
135 West 50th Street, New York, N.Y. 10020

Published simultaneously in Canada by Van Nostrand Reinhold Ltd.

15 14 13 12 11 10 9 8 7 6 5 4 3 2 1

Library of Congress Cataloging in Publication Data

Gallagher, Paul F.
    Parametric estimating for executives and estimators.

    Includes index.
    1. Engineering–Estimates. I. Title.
TA183.G29 1981   658.1'552   81–1569
ISBN 0-442-23997-1   AACR2

Preface

Estimating is one of the most interesting and essential parts of business, and a thorough knowledge of parametric estimating techniques provides the tools with which to attain amazing results, for the techniques were engineered to enlist management participation, as well as the expertise of specialists without their direct participation and thus provide ready estimates.

A great management advantage can be realized by using these parametric estimating techniques for they develop visibility and elements for planning and controls at the inception of a project. *Parametric Estimating for Executives and Estimators* provides this know-how.

Estimating is a desirable profession — often a training ground for future executives. On entering the estimating field, I found that individuals were often made estimators by merely a change in their payroll status, and being given very little direction.

There was also a scarcity of instruction material on estimating techniques other than for construction work. The first estimating techniques of value for application and training were presented by James R. Crawford in his book, *Estimating, Budgeting and Scheduling,* published in 1944 by the Lockheed Aircraft Corporation. I still treasure a copy.

As a result, parametric estimating techniques evolved to satisfy the demands for reliable and consistent estimates, based on relevant performance. In the course of estimating various programs, some of the techniques that were developed to meet an immediate need were very fruitful and easily applied; and it was soon found necessary to document them for future use. An important file of useful techniques soon developed. I organized these into an estimating manual for my own personal use and the instruction of others.

Consistent effort to develop intelligent estimating techniques finally paid off in a way that I did not believe possible. This book presents the five distinct parametric methods of estimating that developed; and which are applicable to production engineering, implementation, and production. In method 2, I discovered a method of normalizing experienced cost data to be used for estimating new programs. Method 5 was a real breakthrough, making the seemingly impossible easy, providing a complete chain of estimating, standards, and standards reporting.

Parametric Estimating for Executives and Estimators presents the ever-present learning curves which demand an integral position in most estimates, with basics, easy techniques, advanced techniques, and pocket calculator assist routines. This portion recaps the data of many teaching sessions and answers to posed questions on learning-curves.

This book summarizes the techniques for ready reference, and it should prove invaluable to progressive estimators and other executives. It also has the following additional advantages:

1. The methods and techniques provide for the easy development of three estimates to bracket the possible range of performance; thus top management can participate, selecting the most optimum figure.
2. The present state of the art in computers and computer programming is compatible with these techniques. Computer assist presents a new challenge in this most satisfying field.
3. The techniques have engineering status so that one can exercise his best judgment.

My life became much easier after I had developed the proper tools for any project that might come my way. With these methods I have provided estimates to meet a deadline deemed impossible by my associates. I also knew that if, or when, challenged regarding an estimate that I need not be embarassed because the supporting data were available; that it had been used correctly, and the estimate was valid.

PAUL F. GALLAGHER
Estimating and Cost Analysis Consultant

Contents

Parametric Estimating for Executives and Estimators

1
Parametric Estimating

INTRODUCTION

The term parametric, as applied to estimating, denotes determination of the position of the estimate for a new program within the limitations of cost parameters developed by experienced costs on similar programs. This is a simple operation for some types of work, but for others there are many variables that make comparative analysis more complex.

Two quotations in Webster's Unabridged present the idea very well: "Is a quantity within the limits of a stated discussion" (T. F. Weldon); and "Four parameters are necessary to determine an event, namely the three which determine its position and one which determines its time" (P. W. Bridgman).

The parametric estimating methods presented in this book have been found very rewarding, providing almost immediate estimates and useful data for many types of assignments. They can make estimating a joyous adventure, for one can state the results of his estimate with confidence.

Three of the five estimating methods presented are applicable at the cost (or price) level or for estimating the hourly requirements. The other two methods deal with cost (or price) only.

It is advisable to make estimates for the total cost (or price) plus a more detailed labor and material estimate when even a rough material list can be realistically priced, because the material costs for many types of work are in the range of 40% to 60% of the total cost. This double check is very important.

CONTEMPORARY ORIENTATION

The three parameters of position and the one of time can be summarized in the term contemporary orientation, which results in three questions:

Where have we been?
Where are we going?
Where are we now?

The prices of the planes in Fig. 1-1 indicate the critical trends that can overwhelm an industry. Each new family of planes indicates a price increase of at least three to one. These factors and dollars are not adjusted for inflation, so we can reasonably adjust them to two to one. Even this type of progression in costs can soon reach a ceiling, and the price becomes prohibitive. The B–70 was not produced in quantities, and every plane that has been started in its place has run into financial difficulty, bumping into the cost ceiling. This trend, a doubling of costs for each new type has carried over into the present time; note the reports that the Concorde costs are twice the costs of the Boeing 747.

Our grandfathers practiced this type of progression for an amusing parlor problem: The question would be asked, "Which would you rather have, five dollars to shoe a pony, or one cent for the first nail, two cents for the second, four cents for the third; this doubling to continue through all twenty-four nails?"

The second choice provides $5.12 for the tenth nail only, $83,886 for the twenty-fourth nail, and $167,772 in total.

Parametric estimating methods are necessary for dealing with such dynamic conditons, to establish the proper bounds within which it is possible for the estimator to use his judgment in a consistent manner.

Dollars may not be a means of measuring progress; but they are an indicator, as they increase, that something is different, larger, more complex, or may use more costly materials and processes. Real progress often brings a reduction in costs. This takes skill, and that which is known as a breakthrough. Although there are many breakthroughs that reduce costs per function, they often do not offset the added costs incurred by the additional demands.

There is very often a lag in technology; the anticipated breakthrough does not materialize as anticipated. Thus, there is a grave business risk

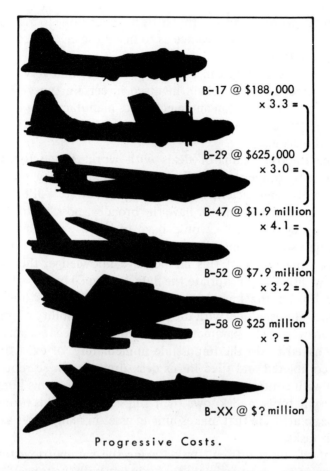

B-17 @ $188,000
x 3.3 =

B-29 @ $625,000
x 3.0 =

B-47 @ $1.9 million
x 4.1 =

B-52 @ $7.9 million
x 3.2 =

B-58 @ $25 million
x ? =

B-XX @ $? million

Progressive Costs.

Fig. 1-1. There seems to be some amazement that the estimated cost for our latest bomber is just over $100 million. However, this is not out of line with the previous trend. If the B-70 had cost $50 million each, then according to past trends it could be expected that the next generation plane would cost $100 million each.

involved when it is assumed that certain items in the laboratory will be practical and at hand to fill a major function. Just one breakthrough that fails to develop can be disastrous; for example: When electric cars were first produced at the turn of the century, it seemed logical to expect a much more efficient battery to be available by the time the first electric cars were to be on the road. However, Mr. Edison did such a thorough search for the proper materials initially, and his engineering was so advanced, that little has been done to make a break-through of sufficient importance as to make the electric car practical even at this late date.

Parametric estimating can be very challenging and very rewarding, even awesome; for the dollars involved in certain types of business such as aircraft production and electronics manufacturing can be very impressive, especially when the production quantities are in the hundreds or thousands.

The parametric estimator deals with work related to the latest scientific developments, meeting and communicating with highly skilled engineers and scientists as part of his routine. All this requires that the parametric estimator have the broadest possible background. To be fully equipped, he would have to know every facet of engineering, industrial engineering, accounting, and all phases of production. Lacking much of this all-inclusive knowledge, the estimator must know how to appropriate the data developed by experts in many fields to bolster the knowledge that he has, for he is out in front (sometimes almost alone) with the clear-cut responsibility of producing reliable answers. The parametric estimator must become accustomed to being asked to do the impossible immediately, or even to have it already completed and filed in his desk drawer. These requests can be faced with confidence and assurance if he has done his homework, has the right tools, and knows their full capacity. This promotes an enthusiastic attitude that makes him an asset in many ways as he performs his tasks.

One of the most difficult tasks facing the parametric estimator is the selling of a correct estimate that is in line with the type of program required, because a budget that bought 100 items last week may buy only six items 10 years later. These ever-increasing costs require that the estimate be consistent with costs experienced on previous programs. This is accomplished by having a firm engineered base, one that can be dollarized by the use of formulas developed by regression analysis of cost data from previous programs. Empirical formulas are also necessary to promote consistency when there is no other answer.

There is no crystal ball, and the multiplicity of the required data is limited by available information and most often must be taken at face value (perfect data are scarce), but those producing and recording it are professionals doing their best.

Many items in this book will mean little to the parametric estimator until he meets a problem requiring some specific subject covered.

Problems of Comparative Estimating

At first consideration it may seem very easy to compare a new item or program to another that has been produced for a certain cost, and thereby assume a very likely cost for the new program. However, this is not necessarily the case.

For instance, if a certain trunk no. 1 were full of money and the total dollars were known, how much money would there be in another trunk, no. 2, of the same dimensions? To make any reasonable estimate, it would be necessary to know the heights of the stacks of bills by denomination in the respective trunks and the total weights for each type of coin. This illustrates one of the major problems and covers only the cost variances due to the mix of the types of components used.

The labor required to count the money and put it in the trunk is another variable that must be part of the total value involved in furnishing the trunk complete. If the trunk no. 1 project were for three trunks to be filled with money and the total labor cost were divided by three to arrive at the labor cost per trunk, would the resultant amount be sufficient labor cost for our new project of trunk no. 2 if the request for quotation called for one trunk only? Would it be too much if the new program were for ten trunks of money? Would there be a difference in labor if all ten trunks were to be filled in 1 month or in 5 years?

If the new project were for 500 trunks of money in 10 months, would we have to develop some mechanical assist techniques that would cut the labor requirements; if so, then what would these costs be?

This simple analogy illustrates some of the problems found in comparing apparently like items. To this can be added the problems attendant with the unique program, and we have the following five elements that affect cost: (1) mix of components, (2) unique components, (3) the unique program, (4) rate of production, and (5) quantity to be produced.

Cost-sensitive Elements. If we are to relate one project to another we must use some type of logical comparison technique. Costs per pound are possibly one of the most universal methods used for comparative purposes, but one must compare items that are made of like components. It is therefore necessary to divide the item being estimated into units for which a realistic and logical comparison can be made. An engineer may use the expression "complexity factor," which is his judgment relative to the products being compared, but this statement does not necessarily have any consistent relation to production costs. It is more apt to reflect the amount of engineering required. Cost indexes or cost ratings are possibly better terms, for we are endeavoring to establish a total cost for the new program.

An estimate is considered good if it has secured the contract and if production costs have resulted in a reasonable profit, but this is an after-the-fact judgment, whereas the estimator is endeavoring to forecast this happy state. It is therefore necessary to be realistic, but even this is often a matter of opinion. One executive that I reported to always said, "Our estimates are correct." This position is good in many ways, except that the estimator may sense that it is his responsibility alone to choose that one right estimate from among the many wrong ones.

The most practical position is to assume that no one estimate is correct, but that a range of costs is possible due to possible variances in the forecast conditions and ground rules. Then the most probable forecast of conditions is made and weighed against the business risks; this is a management operation used to determine the agreed price within the estimated range. These determining assumptions should be documented because excess profits as well as losses must be explained. This book on parametric estimating highly recommends the use of a range of possible costs defined by: target, possible minimum, and possible maximum.

FACETS OF PARAMETRIC ESTIMATING

The elements, data, and techniques used in parametric estimating can, and should be, kept simple to promote accuracy. These are:

1. *Data Banks:* History on similar items. These data must be the same type of elements used in the various parametric estimating methods. (See Fig. 1-2)

2. *New-product Definition:* This must also be in terms that apply to the elements used in the parametric estimating methods.

3. *Five Parametric Estimating Methods:*

Method 1   System Parameters—The Sleeve of Experience (See Chapter 2)
(Learning-curve techniques are required) (See Figs. 1-3 and 1-5)

Method 2   Unit or Function Parameters (See Chapter 3)
(Learning-curve techniques are required)

Method 3   Parameters for Budgeting Men, Materials, and Money (See Chapter 4)
(The use of distribution curves may be required)

Method 4   Parameters by type of Work and/or Material (See Chapter 5)
(Variable budget techniques required)

Method 5   Parameters for Modular Work Measurement (See Chapter 6)
(MTM summarized into modules for a complete estimating and performance reporting chain)

DATA BANK

Programs		Mid–pt. Date	Wt.	Vol.	Prod. Rate	Qty.	Cost
No.	Name						
. . . .	. . . . . .	. . . .	. . .	. . .	. . . .	. . .	. . . .
. . . .	. . . . . .	. . . .	. . .	. . .	. . . .	. . .	. . . .
. . . .	. . . . . .	. . . .	. . .	. . .	. . . .	. . .	. . . .

Fig. 1-2. Key elements of a data bank.

LEARNING-CURVES / COST CURVES

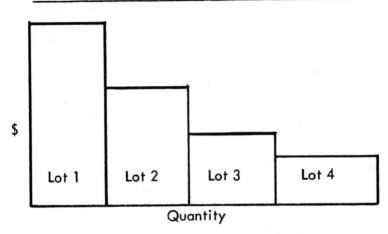

Fig. 1-3. Recorded experienced curves: learning curves/cost curves.

4. *Probability Techniques:* To enlist the full participation of management in the estimating adventure, and thus to have management determine the best estimate for the forecast conditions. This engenders a more personal interest in areas that can make a great difference in the resulting costs. (See Fig. 1-4 and Chapter 7)

DATA BANKS AND PROGRAM TYPES

The purpose of a data bank is to provide applicable parameters and comparable values from previous programs to be used in estimating new programs. Therefore the data bank should contain the same type of data as that used to define a new program for the application of the respective parametric estimating methods.

The best data bank is composed of standard operating reports that have been consistent for a period of at least several years. Best results are obtained by comparing prototype work to prototype work costs. These items are made using hand tools and techniques for the most part. Preproduction work should be compared to experienced preproduction costs as such items are made by hand tools and some production tools that may fit the tasks. Production work should be compared to costs of production work produced on hard tools. Semiautomatic tools, when used extensively, require a fourth class of costs.

PROBABILITY TECHNIQUES

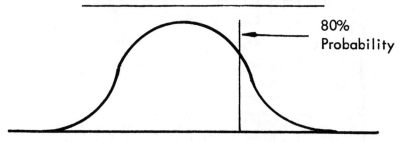

80%
Probability

Fig. 1-4. Safety and probability: probability techniques.

These various phases of production are different, but they are related as shown in Fig. 1-5. The last prototype item can be considered equivalent to the first preproduction item, and the last preprod equivalent to the first production item. These data develop curves when plotted on arithmetic grid paper, but when plotted on log-log paper they can be represented for the most part by straight lines.

A unique system called "learning-curve techniques" provides a method whereby experienced data can be applied to new programs. Figures 1-5 and 1-6 compare the two methods of plotting the same data. Learning-curve techniques are so important, and so often misused, that Chapters 8, 9, and 10 are devoted to them.

Data Banks for Parametric Estimating

The following data have been found very useful in the development of parametric estimating formulas and models. Each datum element should be listed by the total product unit, functional subsystem, and units of the type involved.

In making a list of the required data, it is well to consider the values used in the sales proposal specifications, plus any others that are available at the conceptual stage of a program, such as weight, volume, capacity—whatever it may be, such as input or output; number of circuits and voltages; or horsepower, speed, ceiling, seating capacity, etc. Parts and components lists with the quantity of each type and the state of the art, plus other material quantities are needed. Then there are the various values that are used for control purposes during production such as total raw material per department per month.

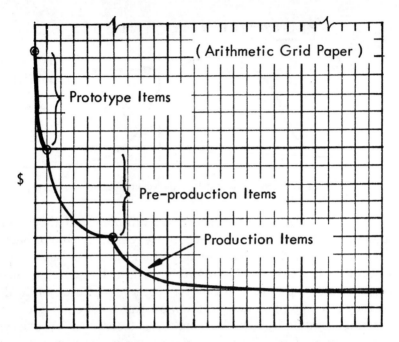

Fig. 1-5. Classes of work—how related when plotted on arithmetic grid paper.

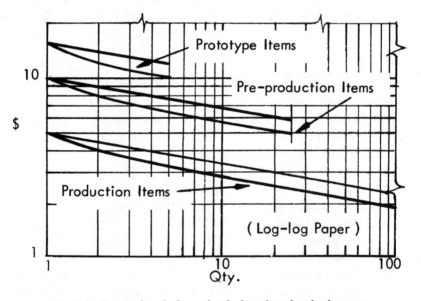

Fig. 1-6. Classes of work—how related when plotted on log-log paper.

Labor control data such as standard hours per item (developed by some standard or uniform estimating value method) are needed, plus a record of actual hours per standard hour from unit one to the completion of the product, with preproduction separated from production. These should be summarized by type of work such as: mechanical, sheet metal, electronic and/or electrical, and quality control; even by cost center if possible.

Much or all of these data are usually available in the modern engineering and production facility, but the task of locating and cataloging them requires special effort preliminary to setting up a computerized data bank. Several notebooks and chart books will suffice for some time.

One can start by locating as much consistent, pertinent data as possible and then determining what more is necessary. This too may then be found.

Data on parts also will be found necessary for the development of the needed charts and formulas for method 5.

NEW PRODUCT DEFINITION

Merits of an Engineered Base

The engineered base makes it possible to refer back to a definite starting point; additions and deletions can be identified for what they are. A change of concept for the program will be noted as such and not as an overrun or an underrun, as either one can imply that a poor estimate was made or that someone was negligent. The following incident illustrates the problem well:

An estimator was asked to make an estimate that was immediately needed for an electronics box based on a component count and a general description of the box. These simple details represented the engineer's conception of the box at that date. He used these data with method 5 (presented later in this book) to develop his estimate. In about an hour he had completed the estimate for the implementation and production of the box and handed it to his supervisor.

A couple of months later the supervisor walked up to the estimator's desk with a disappointed mien, desiring to know what was wrong with "that estimate"; for the box had cost less than estimated. The estimator pulled a copy of the estimate from his drawer to compare the estimating data with a bill of material actually used. The quantity requirements for one of the major components was six on the bill of material, but twelve on the engineer's component count list. The responsible engineer was immediately called for an explanation, which was: At the time of the initial request there was a possibility that the box could be engineered with six of the components, but he knew that it could be done with twelve. He therefore had no choice but to call for twelve. It later developed that six components could be made to do. The revised quantity of components was used to recalculate the estimate, and the new result was very close to the actual costs. As a result, the supervisor was happy because the estimating method was again confirmed, the estimator had performed his task properly and he was happy, and the credibility of the engineer had been verified. All this would not have been possible without the definite engineering base.

If the engineer had run into unforeseen problems and had had to use more components than initially estimated, the explanation would have been the same, and it would not have reflected badly even on the engineer, for he had used the best information available at the time.

Much of the recognition that all of us desire and need for a happy life depends on the credibility of our work, and we must be as free as possible from worry to perform our duties in an efficient manner. This too, can be achieved only through an engineered approach to the various estimating problems.

Was It an Overrun or a Change of Concept?   The importance of interpreting the actual costs resulting from firm engineering versus the costs dictated by the basic concept cannot be overemphasized. If this simple process could be a required procedure, there would be fewer headlines in the news stating that some program was costing much more than initially calculated. The cost analyst sometimes encounters programs that were publicized as gross overruns, when the facts show that the statement of work had substantially changed from the initial concept and that these changes had not been properly accounted for by the reporter.

Two Levels of Estimating

Parametric estimating can be at the total cost (or price) level, or the estimate can be built on a detailed material and labor estimate which is then priced. Cost level estimating is usually considered not as reliable as building up the costs by labor and material details, because there are so many variables that affect the costs, and some are difficult to deal with at the total cost level; also, the customer is usually not satisfied with anything less than a detailed breakdown of the costs.

The Basic Information

Selecting the right data before starting the estimate is extremely important, for the estimate can do no more than translate these data into dollars.

The Official Description. One should secure all of the data that can be considered as part of the official description. This data package may vary from project to project. Seldom can the parametric estimator find the full definition that he needs unless he can make the data mandatory at his request. It is advisable to itemize the needed types of data as soon as possible; this list should be provided to all parties active in supplying these data so that all can work efficiently in developing uniform data packages.

Request for Quotation (RFQ) Outline. The RFQ usually includes many elements that are not pertinent to the development of the estimate, so it is not practical to use in its original form, but must be reduced to its simplest requirements for ready reference and a full grasp of the responsibilities of the parametric estimator. A yellow marker or grease pencil can be used to mark right across the printing of these pertinent items; it will make each item very easy to find when needed as each will appear as if illuminated. A summary of these items can also be made so that the estimator will be sure to cover every item as requested. Great importance is placed on being responsive in every particular.

Pertinent administrative guidelines and decisions in regard to the program should also supplement this outline. A great deal of effort is often wasted because this phase of the work is not given the proper priority.

The present trend is in the direction of unique programs. These cause the most concern because they are difficult to relate to past programs, but even unique programs can be more easily estimated if they are properly defined and handled in a systematic order.

Conceptual Engineering Conference Output. Parametric estimators soon note that there can be much improvement in the engineered base, and that there is a need to outline and set up a corrective procedure and have everyone follow it if the many parametric estimating problems are to be dealt with successfully. A profitable procedure can be developed by making a composite plan, such as follows, which includes the fruitful activities and outputs of past successful conferences.

Much of the data package, except for the RFQ, some of the schedules, and a few minor elements, must be developed by the responsible engineers. This requires an engineering cadre to meet for the purpose of formulating the design strategy and materializing their thoughts into data that can be communicated to the customer and to all estimating and concerned production personnel. It is seldom that the full potential of such an informed and capable group of engineers is realized because they do not know exactly what is expected of them.

There is an interrelationship between certain capabilities that can be used to great advantage in developing a firm engineered base for an estimate. The parametric estimator (or cost engineer) cannot operate alone, independent of the responsible engineering groups. Even if the estimator possessed designing capabilities equivalent to those of all of the responsible engineers and visualized an ideal design, it would be of little use in developing an estimate, because the estimate must reflect the product design conceived by the responsible engineer within the specific design restraints given to him. This is one problem requiring engineering know-how, for formulas have not yet been developed to supplant the engineer.

Full use of all capabilities is required, and it has been proved by many that the interaction of the various capabilities and techniques will produce well-evaluated results in a minimum amount of time. Each of the following specialists is incomplete and ineffectual without the others: The program manager is needed to outline and maintain responsive limitations. The project engineer's duty is to maintain system integrity, and subsystem specialists are required to propose and assist in the selection of optimum subelements and describe them effectually, using rough sketches and listing the proposed type of components and their quantity requirements. Production illustrators should be included in the definition cadre to clean up the rough sketches and illustrate each task, unit, piece of test equipment, etc.

A firm base can be quickly developed if these capabilities have their tasks outlined and each understands how his work complements that of the others. The goal should be to define and/or promote: (1) the big picture, (2) the task picture, (3) component requirements, (4) a review by the cadre, and (5) review by top management.

The development of each picture (box, task, or phase) requires: discussion, rough sketches, and notes; preliminary sketches by the illustrators; the hanging of the sketches in the proper sequence on the wall for all those participating to see and constructively criticize; then review and correction to satisfy all levels of management having assigned responsibility.

This approach to defining a system will proceed at a very rapid pace if it is understood by all that the task has a deadline. Several engineers and a couple of production illustrators who are apt at sketching on 30" × 40" paper can cover much of the necessary work of defining a system in a few days. The reviews and revisions can be intelligently done in one day if necessary. The approved data sheets can then be photographed and reproduced on 8.5" × 11" sheets for immediate reference and as a definitive base for later reference. This procedure can do much to preserve the integrity and credibility of an organization.

The purpose of this book is to present practical parametric techniques and engineering methods to develop satisfactory models that are representative of a hypothetical firm rather than to supply set formulas that will produce estimates that are applicable for any firm operating with its unique capabilities and restrictions in any locality. Such an attempt at producing universal models could only be presumptuous and misleading. However, it should not be difficult to revise and adjust these models to make them compatible with the unique needs of many comparable firms. It is hoped that these models and methods will demonstrate their feasibility and, encourage estimators to pursue and develop their own models to cope with their own unique requirements.

Justification for the Documentation of the Ground Rules and Assumptions. An estimate is no better than the documentation of its ground rules and assumptions. For instance, a request for quotation was received by a firm for an item that was then in production at a low point on the learning curve, over a thousand of the items having been produced. However, the time of the delivery schedule in the RFQ was about a year after the line was scheduled to be torn down to make room for another project. Therefore the ground rules given the estimator were: (1) The project requested would be produced at a new facility. (2) New workers would be used. (3) There would be about a year between the end of the old contract and the start of the new one.

The estimate was made to conform to these guidelines, and even the customer agreed that the price was right. However, the contract was received in an unusually short time with a schedule that tied into the production on hand. There was no new facility, no new workers, and no lapse in production.

The profit picture of over 60% was alarming until the documentation of the ground rules was pulled from the files. It was then obvious that the estimate was correct; it was the new early schedule that changed the conditions from those forecast, a situation that management could not possibly have foreseen.

This might seem to be an ideal situation, but it was not so, for many hours were spent in an effort to determine what portion of the profit had been earned and could be kept.

Situations such as this are rare, but day-to-day estimating profits in a like manner from the documentation of all ground rules and assumptions.

CONCLUSIONS AND MAJOR POINTS

1. Cost parameters are necessary because:

    1.1 They reduce the scope of the task to the essentials.

    1.2 The impact of forecast conditions can be evaluated and included.

    1.3 All responsible parties can get their viewpoints included.

    1.4 Management can determine the final estimate relative to probability and experience.

2. Data banks must serve the parametric estimating methods. They must therefore be in the same terms as those used to define each new project.

3. One should become familiar with all five methods of parametric estimating and use two or more of them on each project for added proof of accuracy.

4. Like phases of work should be compared where possible; if not, consideration should be given to the possible relationships as shown in Figs. 1-5 and 1-6.

5. A thorough knowledge of the theory and use of learning curves is of major importance, as one of their many uses is to arrive at comparative values relating the projects in the data bank to new projects.

6. The new-product definition must be engineered and fully documented for future reference.

7. The documentation of all ground rules and assumptions is imperative.

2
Method 1: System Parameters—The Sleeve of Experience

The first of the parametric estimating methods deals with the complete project or with the easily identifiable parts or elements. Comparing one item to another, or even to an identical item or program, seems very simple and straightforward, but adjustments must be made for such items as the following:

1. Inflation or deflation for the difference in time.
2. The effect of total quantity produced on material cost each and the average hours each.
3. The rate of production having a great affect on the rate of cost improvement.
4. Whether the workers are familiar with the type of program.
5. Whether the tools, test equipment, and facilities are the same class as used on the projects from which the data were taken.
6. How the new program will rate in priority.
7. The state of the art in the production of key components, a very critical element during the last 20 years. Components that have been on the market for only one year at $110 each may be only $5 each 5 years later; and parametric estimates often lead production and purchasing by 3 to 5 years.

All of these problems as listed can be dealt with in a realistic manner as presented in the following cases. Some problems are handled directly, while others are included by judgment factors from the responsible groups and management.

Inflation of deflation for the difference in time: The causes of the yearly inflation of costs are not necessarily uniform in all areas or types of production, some being affected much more than others; thus, all available indexes may be wrong for direct use for the industry being considered. So for lack of a crystal ball, one must resort to some other method such as that used in Chapter 3, where the inflation is assumed arbitrarily to be 6% a year; this is then added to the historical data to make all data consistent dollarwise. These data, when further adjusted for density and newness, still indicate an upward slope of 1.02% a year for the major portion of the units considered. This is added to the 6%, resulting in an inflation index of 7.02% per year for the experience data used, and it is assumed that the future offers nothing to reduce the forecast rate to a lower figure.

The effect of total quantity produced on material cost each and the average hours each: Cheaper for larger quantities, whether by the dozen, hundreds, or thousands is much the rule. This can be considered a cost-reduction curve using learning-curve techniques for general application; see Chapters 8 and 9.

Labor costs for ever-increasing quantities have long been interpreted by learning curves and their various techniques. The cost-reduction elements for material and labor must be considered in light of the prevailing market and past experience.

The rate of production having much to do with the labor costs: A very light schedule will often show but little improvement after the initial assigning of men to each station. Whereas a reasonable increase in the rate of production per month will normally show much improvement, too high an increase in the scheduled rate can cause much confusion in all production areas and add to costs.

Advantage of having experienced workers: Workers who are familiar with the type of work involved in a new program will produce the initial items most efficiently; it will be almost like a follow-on of the previous product.

The type of tools, test equipment, and facilities: Reliable estimating data require that similar tools and test equipment be used on the products from which the data originated as are planned for the new program; otherwise some proper allowance should be forecast.

Program priority: The priority given a new program has much to do with costs, especially if only new personnel are assigned to the project and the most experienced supervisors are not available.

Forecasting the state of the art: These forecasts are in the range of 1 to 5 years for parametric estimating. Figure 3-2 (p. 35) shows how critical this can be and provides a possible method of forecasting an adjusting factor. Few firms will furnish even tentative forecasts of critical cost items (those items having great cost-reduction potential as the state of the art matures), for they can be wrong, too. The costs of solid-state electronic items with their ever-increasing miniaturization of circuits are still difficult to forecast. A large market sometimes develops for one type of component and too many local and foreign firms start producing it. As a result, the market is flooded and the price falls drastically. Some estimators then assume that that will be the case in all future projects, but the odds are not there.

Management judgment will always be important. How or why a manager determines any factor or relationship of one program to another is often mysterious, but it is usually his judgment of the consensus of the opinions of others.

THE SLEEVE OF EXPERIENCE IN GENERAL

The "sleeve of experience"* is a very applicable name for this method of parametric estimating because the average costs of a group of comparable programs, when equated or properly weighted and then plotted on log-log paper, dollars versus quantity, usually fall within a certain pattern similar to a sleeve in shape.

Each firm has a unique sleeve of experience into which all of their experience data of completed programs will fall. For some firms this sleeve is very full, and for others it is quite slim. There is always the tendency, however, for one to consider this sleeve as being a line, having no width. This may be desirable, but it is not consistent with the facts, for the conditions relative to each program are often quite different (see Fig. 2-1).

Certain techniques have been developed to apply these experienced data to new, anticipated, or planned programs. This is done through the application of learning-curve techniques. Those unfamiliar with learning curves are referred to Chapters 8, 9, and 10 of this book.

---

*Thanks to James Crawford, who used this description in one of our conferences.

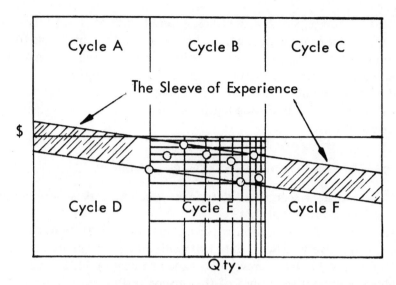

Fig. 2-1. The sleeve of experience.

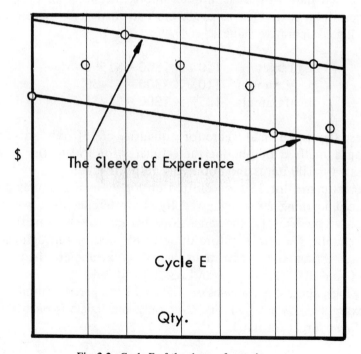

Fig. 2-2. Cycle E of the sleeve of experience.

Logarithmic paper, 2 × 3 cycles, printed on 8.5″ × 11″ paper, is one of the basic tools needed. It has been used in this book for all learning-curve work. However, the text format being minimal, it is often necessary to use only one cycle, as shown in Fig. 2-2.

It is recommended that the reader diagram each illustration on a full sheet of 2 × 3 cycle log-log paper; this is always good learning-curve procedure.

The Sleeve of Experience from Data on One Program Only

There is often a scarcity of data from which to work, so an example of this type is presented first.

The request for quotation is for a quantity of 100 items and 300 items, respectively. Useful data from only one program are on hand: for 150 items that were previously made at a cost of $1300 each.

Three comparative ratings of the proposed program relative to the experienced data will assist in determining the sleeve of experience. The best informed, such as engineers and managers, agreed on the following comparative values:

a. Target   120% × 1300 = $1560
b. Minimum   110% × 1300 = $1430
c. Maximum   135% × 1300 = $1755

These figures (estimates) are for a quantity of 150 only, so we must develop a method of relating the estimates for 150 to the required forecasts for 100 items and 300 items, respectively.

The estimates for 150 are plotted on log-log paper, and as the experienced learning curve followed by the reference data was on an 89% curve, we can use the same slope because of the sensitive data being equal. We can therefore draw an 89% curve through each of the three estimates for 150 items, and thus extrapolate the costs for the 100 and 300 quantities as in Figs. 2-3 and 2-4.

Log-log paper, 2 × 3 cycles on 8.5″ × 11″ paper, was used to plot this example; however, in Fig. 2-4, only one cycle is needed, thus providing a larger format.

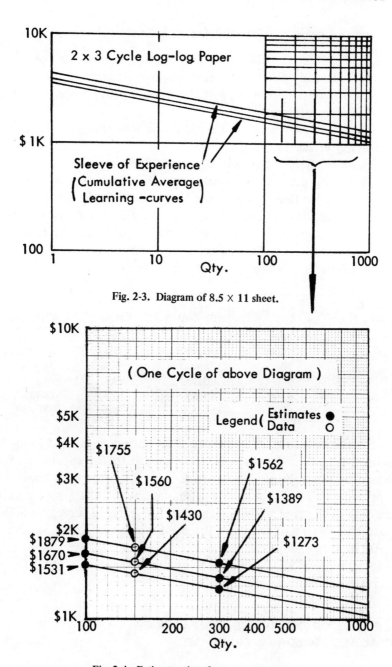

Fig. 2-3. Diagram of 8.5 × 11 sheet.

Fig. 2-4. Estimate—data from one program.

The Sleeve of Experience — Several Reference Programs and Average Costs

The procedure as outlined is a very good method to use in conference. The study deals with the direct construction of the sleeve of experience from a number of initial programs only, posted at the last unit of each program.

Problem: The request for quotation defines the ground rules for an estimate on 700 items.

Procedure: Select a group of comparable programs and tabulate the initial program quantities and costs that have been adjusted for the anticipated inflation and other time elements. Post these figures on a large sheet of paper in columns 1, 2, and 3. Column 4 is used to tabulate the comparative percent estimates made by the respective conferees. These percentages are then used as factors of the basic data of column 3 to arrive at the estimates in column 5.

1 Proj. No.	2 Qty.	3 Adj. Cost	4 Est. %	5 Estimate
1	100	239,000	127	304,000
2	150	400,000	80	320,000
3	200	358,000	95	340,000
4	400	90,000	133	240,000
5	530	160,000	150	240,000
6	800	350,000	80	280,000

For conference use, a large sheet of layout paper is ruled in log-log fashion for posting the column 5 estimates at their respective quantities as noted in column 2. Figures 2-5 and 2-6 show how these data are posted, and how the sleeve of experience is made by enclosing these data in the typical cumulative average learning curves. It is then a matter of judgment to select three estimates that are applicable to the 700 quantity.

The estimate is $255,000, the minimum estimate is $220,000, and the maximum estimate is $285,000.

THE SLEEVE OF EXPERIENCE FROM A MIX OF INITIAL AND FOLLOW-ON PROGRAMS

When the basic data are from a mix of follow-on programs and initial programs, the system must be modified in that all data estimates must

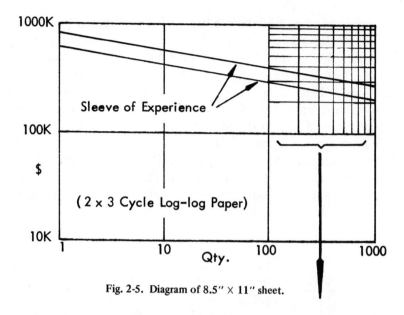

Fig. 2-5. Diagram of 8.5″ × 11″ sheet.

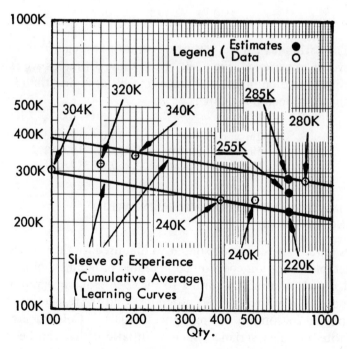

Fig. 2-6. Estimate—several reference programs.

be plotted at the unit of average value of the respective programs; this develops the learning-curve unit line rather than the cumulative average line used in the previous exercise. The data routine is the same except that, for the follow-on programs, the first and last unit (serial numbers) are required; these data are to be posted on log-log paper as follows: For rough estimates the average value for an initial program is near the unit value of 33% of the quantity on the unit learning curve. The unit of average value for a follow-on program is near the unit at the arithmetic midpoint of the program on the unit line, Figs. 2-7 and 2-8.

Problem: An estimate is required for 75 items.

Procedure: Comparative data are available for three programs; the first and the third are follow-on programs, and the second is an initial program.

Proj. No.	Qty.	Serial Nos.	Mid. Pt.	Adj. Costs	Est. %	Estimates
1	170	10–180	95	338	65	220
2	45		15	309	110	340
3	125	25–150	87	225	80	180

The estimates for the seventy-fifth unit are read from the resulting sleeve of experience, Fig. 2-8. Then these estimates are raised to the learning-curve cumulative average line by applying the $(1 - K)$ value of the developed 89% slope.

Target estimate	230/0.8319 = $276
Minimum estimate	185/0.8319 = $223
Maximum estimate	260/0.8319 = $313

ESTIMATING THE COST OF IMPLEMENTATION USING METHOD 1

This requires a technique similar to that used for the product estimate. All implementation costs on comparable programs should be retrieved and noted on a matrix similar to Fig. 2-9. Each program title should be noted under "referenced program," and the respective costs should be posted in column 1, "reference actuals." Column 2 can be used to note any cost variance, such as inflation, that can be determined for the proposed program in its scheduled period as related to that

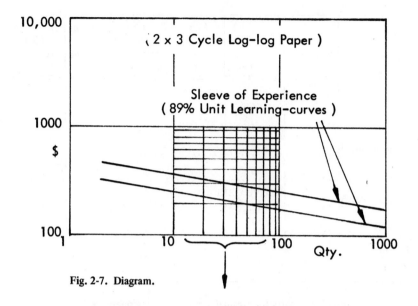

Fig. 2-7. Diagram.

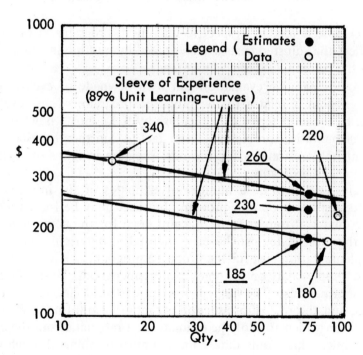

Fig. 2-8. Estimate—follow-on data included.

Referenced Program	Rate Per Month	(1) Reference Actuals	(2) Inflation Factor	(3) (1) x (2) Adjusted Ref. Data	(4) Relative Factor	(5) (3) x (4) Estimate for Noted Rates
Program No. 1						
Program No. 2						
Program No. 3						
Program No. 4						
Program No. 5						
Program No. 6						
Program No. 7						

Fig. 2-9. Matrix for organizing and processing reference data.

experienced for the respective referenced programs. Column 3 can reflect the adjusted costs as determined by multiplying column 1 data by the column 2 factor.

A comparative evaluation can then be made of each referenced program as it relates to the proposed program. These evaluations can be made in terms of percentage by those most knowledgeable of implementation problems. These factors can then be used to make comparable estimates for the new program in terms of each of the referenced programs.

This will provide a number of estimates of the implementation costs for the proposed program, but each may be for a production rate that is different from that proposed for the new program. All of these estimates can be used in total to determine the estimated implementation cost for the proposed program.

A simple chart on arithmetic grid paper should be satisfactory for most situations. The implementation cost in dollars can form the Y axis, and the rate of production per month can form the X axis for a chart on which to plot the estimates and rate data from the previous schedule. This should determine a number of points through which a line of best fit can be drawn by inspection (see Fig. 2-10).

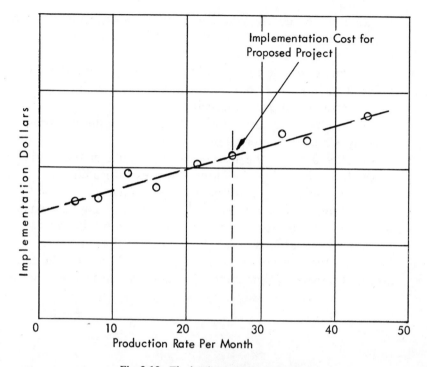

Fig. 2-10. The implementation estimate.

We can then read the estimated implementation cost for the new program at the planned rate of production where the line of best fit crosses a vertical line drawn through the planned production rate for the new program. This provides a practical method of making use of cost data banks.

CONCLUSIONS AND MAJOR POINTS

1. When all of the data deal with initial programs, the cumulative average figures are used; these are plotted at the last unit position, thus developing the cumulative average learning curves.
2. For a mix of data, initial programs and follow-on programs, these data are plotted at the unit of average value to determine the unit lines. The last unit of the estimate can then be read from these unit lines. Then the cumulative average for the estimated

program can be developed by dividing these last unit figures by the learning-curve $(1 - K)$ value that is applicable to the experienced slope.

3. If it seems desirable to use only one method, then the system referred to in item 2 above for a mix of data can be used in all cases if accuracy proves unimpaired.

4. The typical learning-curve slope experienced on the data provides a ready slope for determining the sleeve of experience. Processing the data by the least-squares formula or the averaging method to determine the slope provides misleading results in this type of estimate, from these kinds of data.

5. This method of estimating is very good for obtaining a fast answer by the individual estimator, or in conference, but one or two of the other five methods should be used to confirm it.

6. Note that the range of costs is provided in all cases by the calculation of target, possible minimum, and possible maximum costs.

3
Method 2: Unit or Function Parameters

This method approaches the ideal in that it is applicable to functions just as well as it is to units or subsystems. It is built around the simplest principle: that anything can be estimated if one knows its weight and has a reliable cost per pound multiplier. The idea is also applicable to any measurable function.

There is a base figure that remains after all disturbing elements are removed from historical costs per pound or costs per function index. After establishing the base, or constant, we can forecast the elements that may distort the cost per pound (or function) and include them as factors to the base, and thus we obtain a model (formula) for estimating.

In the development of the constant from historical data, the following operations are applied to each type of data, by each unit or each specific function. Program quantities vary, so all data must be interpolated or extrapolated to a common quantity on the cost reduction curve; the average cost of the first 1000 units is used here. The cost per pound (or cost per function index) for each historical program must be calculated, and such data often cover a number of years, so older data must be adjusted to current dollar values.

There may be other variables that must be removed to provide a usable constant. Semilog paper with a vertical logarithmic scale is ideal for dealing with these other variables, because the factoring out of any distorting elements results in a more linear plotting. Overcorrection or undercorrection can therefore be judged. This provides a method of dealing with every known variable, using all available information. Figure 3-1 is a typical example of developing a constant for an airborne computer.

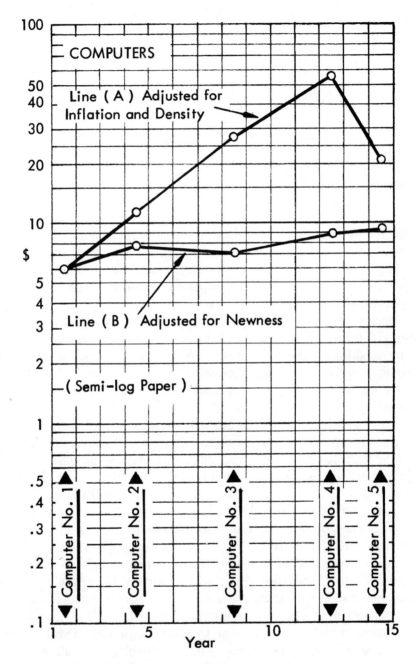

Fig. 3-1. Developing the constant on semilog paper.

DEVELOPING THE CONSTANT

The process used in developing the constant is illustrated by the processing of some airborne computer data. Figure 3-1 shows the data results after adjustment for inflation and density (these being simple routines) as shown in line A.

The typical procedure uses semilog paper, the vertical scale being used for cost per pound (or unit of measure per function) and the horizontal scale covering the years spanned by the historical data.

A plot is made of each historical program at the midpoint of its production period relative to the yearly scale and its cost per pound (or cost per unit of function) relative to the vertical scale. A line is then drawn to connect all the plotted points; it will probably be very irregular, but it can be straightened by removing the variable elements.

Density is often a variable element that must be dealt with. The weight divided by the cubic feet of volume is a good factor to experiment with. If it overcorrects or undercorrects, one can adjust the numerator or denominator until it produces results that seem reasonable for all items. The line should become less irregular when the density element is removed. (Other indexes of density are good when available.)

For some types of work, removal of the density element will provide sufficient linearity to the plots to establish the constant and the model, but not so with computer electronics and other dynamic industries.

Newness of components and techniques creates the greatest variance, so a method is needed whereby these elements can be dealt with in a consistent manner. Line B in Fig. 3-1 shows the results of applying such a method. The results are sufficiently linear to establish the constant and to determine the model.

A SOLUTION TO THE NEWNESS AND STATE-OF-THE-ART COST PROBLEMS

The newness element can be delineated by determining:

1. The percent of the item composed of relatively new components and the length of time that they have been on the market.
2. The percent of the item requiring the use of new techniques and the length of time that they have been used.

A simple nomogram such as Fig. 3-2, combining both new components and techniques, can be developed to provide the newness factors on the index scale (*N*). Scale *A* is used to represent the percent of the product affected; it is used in conjunction with, and parallel to, the time scale (*Y*), which is used to show how long the components or techniques have been used. Note that the percent scale decreases toward the bottom of the page, but the years increase.

Experience indicates that in 7 years the cost of components and the premium on new techniques levels out; this establishes the span of the time scale. New types of components when first available commercially are at least ten times as high as their final plateau; so this is the span of the index scale. The values of the index scale were graduated so that the leveling-out would be easy. A few trial nomograms were applied to the data, and soon the developed indexes produced near-linear results that could be used in the model.

The resulting line *B* in Fig. 3-1 was still rising at a rate of 1.02% per year, which when added to the arbitrary 6% first used, provides an inflation rate of 7.02%. The fifteenth year was the base for the model, so $9.30 is the constant for an airborne computer. (Constants for other electronics items are presented for a feasibility example only in Fig. 3-8.) All results were then formulated: $D \times W \times N \times C \times I =$ average cost of 1000 items.

MODEL FOR BASIC PRODUCT QUALITY—
AVERAGE COST EACH FOR 1000

The model (formula) uses the following elements:

D — Density (lb divided by ft^3)
W — Weight (lb)
N — Newness factor (from Fig. 3-2)
C — Unit constant (developed from historical data)
I — Inflation factor (as projected from base year of constant)

As these elements were all part of the historical data, adjusted to the average for 1000, they can be used to build up the estimates for future work. Thus: $D \times W \times N \times C \times I =$ average cost each for 1000 items.

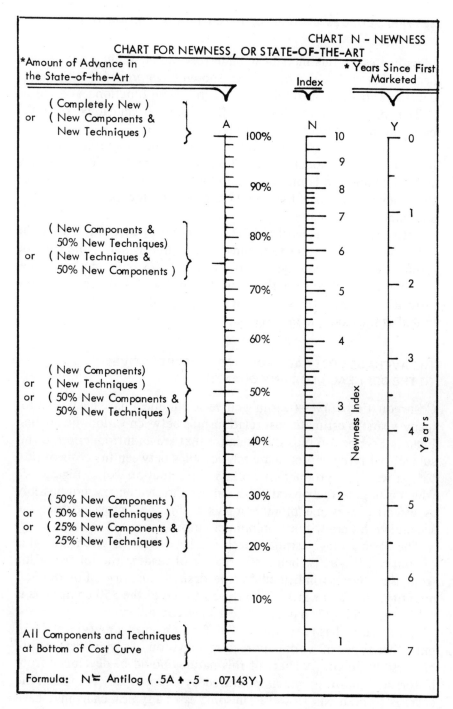

Fig. 3-2. Newness and state-of-the-art index.

Estimates by this model can be in very broad, all-inclusive terms if desired; but visibility and accuracy should be improved if the product constants are developed relative to each system function.

The inflation index should be the estimated change between the base and the midpoint of the production period.

A data sheet (form) for each functional group of a project is an aid in the use of this model. Figure 3-3 is a typical unit (or function) data sheet detailing the computer functions for a project. It provides for three newness indexes that can be used to develop maximum, target, and minimum estimates.

Estimation of a typical airborne computer by method 2, using the data sheet for computer functions, is given in Fig. 3-4. Note that the estimate is for the average for 1000 items. A simple method of developing the estimate for the project quantity of 350 items from the estimate for 1000 items is the next step. The target estimate is presented as the example in this case.

THE AVERAGE COST EACH FOR 1000 ITEMS RELATIVE TO THE COST EACH FOR ANY QUANTITY

Cost-reduction curves (learning-curve techniques) have been used for many years to estimate cost relationships between various quantities. Figures 1-5 and 1-6 plus their relative text are an introduction to this subject, with the approximate relationships between the costs of prototype work, preproduction work, and production work. Figure 3-5, which is used in this operation, is similar and elementary, but possibly as effective as some more complex techniques, provided that the schedule rates are within a minimum range.

The basic average estimate each for 1000 items is given the value of unity (1.0). It is then just a matter of reading the relative value from the chart as indicated by the desired quantity. For the 350 quantity, the factor is 1.17, so the cost each of the 350 computers is 1.17 times $53,772 to equal $62,913 per computer.

A 90% curve has been used in Fig. 3-5 as being near a typical slope, but other types of work may require a revision of several percent up, or possibly down. A chart of this nature should be developed from historical data on programs produced by the company for which the chart is to be used. However, in some cases a typical chart may have to suffice until the proper data can be compiled.

UNIT DATA SHEET

Analyst _Jim Didit_____ Date_____ Ref. No._____ Date_____

Dept. No._____ Engineer _Tom Thunkit_____ Phone Ext._____

Project _Computer -- Airborne_____

Unit Name or Category _Computer Units -- Misc._____

Production Span --- _1_ / _xx_ to _12_ / _x2_ | Midpoint of Prod. Effort _X1___ |

Total Quantity _350_ Rate Per Month , Max. _35_

Weight | W= _130_ | Size: ___ x ___ x ___ = Volume - Cubic Feet | V = _7.6_____ |

Density Index W / V = | D = _17.11____ |

	Maximum	Target	Minimum
New Components %	30	25	0
New Techniques %	20	17	0
Empirical Average %	25	21	0
Average Years on Market or Years in Use	5.0	6.0	7.0
Newness Indices From Chart N	1.9	1.5	1.0

Fig. 3-3. Method 2 data sheet.

ESTIMATE FORM   Sheet ___ / ___

Project: _Computer -- Airborne_____ Ref. No._____

Analyst : _Jim Didit_____ Date :_____

Total Qty. _350_ Max. Rate / Mo. _35_ Midpoint of Effort: _X1_____

PRODUCT Description or Unit Number	Density Index D	Weight W	Newness Index N	Unit Constant C	Inflation to 1970 Base I =	Subtotal	Cost / Qty. Factor Q	Factory = Cost Each
COMPUTER	17.11	130	1.50	9.30	1.07	#33,201		
PWR. SUPPLY	100.0	40	1.0	1.38	1.07	5,906		
MEMORY	75.0	30	1.33	4.58	1.07	14,665		
SUBTOTAL						#53,772		

Fig. 3-4. Method 2 estimating form.

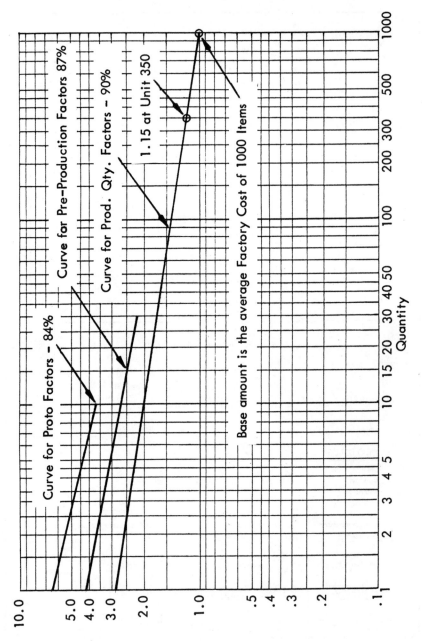

Fig. 3-5. Cost/quantity curves.

This method can be extended to provide combined estimates for the production planning and implementation. Various types of engineering that are related to the program can also be related to production costs at some specific quantity such as the average cost each for 1000 items as used here.

THE IMPLEMENTATION ESTIMATE BY METHOD 2

The relating of implementation costs to the cost of a certain quantity of production items has been used as a rule of thumb by many estimators for some time. It has also been observed that when the implementation capability was doubled, often a certain multiplier was indicated. These two principles have been combined and historical data have been used to formulate this portion of the model also.

The plan is to use the average cost each for 1000 items as the base. This is provided by a subtotal on the project portion of the estimating form. Then the cost to implement for one item per month is the starting point needed to make a chart for any rate per month.

The analysis of the historical data indicated that the quantity of systems equal to the implementation cost for one per month levels out after several years in a particular type of work and can be considered a constant. Our data leveled out at seven equivalent systems' cost being equal to the implementation cost for one item per month.

A curve depicting the increased cost of implementation relative to the rate of production capacity was found to fit the reciprocal to the 74% learning-curve unit line, thus providing a firm definition for this curve. Therefore, the reciprocal to the 74% T_u learning curve (used as a multiplier) will adjust the cost of the implementation for one per month to any rate per month desired.

These two sets of factors can be combined for simplicity and ease of application into one curve and one set of factors as in Fig. 3-6 in which seven systems' (items) cost are equal to the implementation cost for one system (item) per month. This method is given as a feasibility model only.

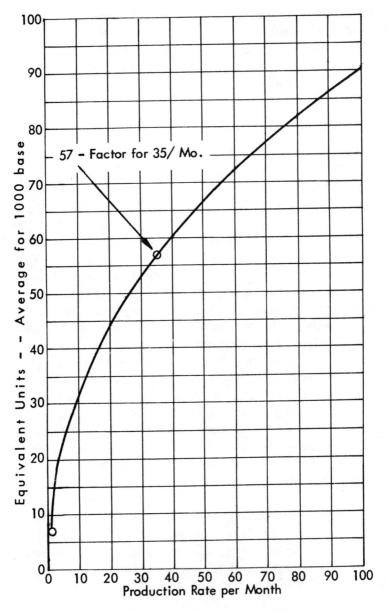

Fig. 3-6. Implementation—equivalent units' cost.

THE COMPLETED ESTIMATE FOR PRODUCT
AND IMPLEMENTATION

A simple estimating form is necessary for the easy application of any estimating system as it reduces the task to a minimum and promotes accuracy in that it includes all the necessary elements. Figure 3-7 completes the estimate for the airborne computer that has been used as a typical project for method 2.

Figure 3-4 carried the work through the base, which is the average cost each for 1000 items. The quantity factor of 1.17 was determined by the cost/quantity curves of Fig. 3-5, thereby providing the cost each for the specified quantity of 350 at $62,913 each. The cost of the implementation was calculated by applying the rate-per-month factor as determined by the chart, Fig. 3-6, to the basic cost; or, the factor of 57, times $53,772 (which is the basic cost), equals $3,065,000 for implementation.

There may be times when it is legal and proper that this implementation figure be reduced, such as when a similar project is in work and certain pieces of implementation have sufficient idle time for a big portion of the new project.

Three different estimates can easily be made by using the three different estimates of newness as given in Fig. 3-3.

I personally developed method 2 in 1968 and first presented it at a conference in March of 1969. Any similarity of this model to any now in use by any firm is coincidental, and merely substantiates the value of all such models.

METHOD 2 IN GENERAL

Even in its present feasibility state, method 2 is very practical in cases requiring one's best judgment for fast, consistent estimates, whether for first or second opinions. Many possible refinements can be made to method 2 as one develops one's own unique version. Experience and/or study in chart work and in the construction of nomograms with their formulas is a requirement for the needed regression analysis used.

METHOD II

ESTIMATE   FORM   Sheet 1 / 1

Project: COMPUTER-- AIRBORNE   Ref. No. _____

Analyst : JIM DIDIT   Date : _____

Total Qty. 350 Max. Rate / Mo. 35   Midpoint of Effort: X 1

PRODUCT Description or Unit Number	Density Index	Weight	Newness Index	Unit Constant	Inflation to 1970 Base	Subtotal	Cost / Qty. Factor	Factory
	D ×	W ×	N ×	C ×	I	= ($/ K$_A$) ×	Q	= Cost Each
COMPUTER	17.11	130	1.50	9.30	1.07	$33,201	1.17	38,845
POWER SUPPLY	100.0	40	1.0	1.38	1.07	5,906	1.17	6,910
MEMORY	75.0	30	1.33	4.58	1.07	14,665	1.17	17,158

PRODUCT --Total Factory Cost Each ------------------------- 62,913

IMPLEMENTATION

(1)   Average Cost Each for 1000 (Subtotal) $53,772

(2)   Cost / Rate-of-Prod. Factor   57

(1) × (2) = Implementation Cost for $3,065,000
      a Production Rate of 35 Per Mo.

Fig. 3-7. Method 2—completed estimate.

This method is one that should fit the constant future changes as they occur, for the following volatile experience data of many years were easily evaluated and made to concur in this model. When these experienced costs of dollars per pound were plotted, they developed curves as in Fig. 3-9. These curves were not practical for pricing, for rather than one curve, the data developed a family of curves, each one applicable to a singular type of construction or a particular mix of components. It will be noticed in Fig. 3-10 that there are five different families of products shown on the curve, each being more complex or more costly than the previous. Various programs, A, B, and C in Fig. 3-10, are all in the same family. Then a small amount of miniaturization was introduced in program (D), thus introducing family number 2. Greater and greater miniaturization developed families 3, 4, and 5. This shows that the cost per pound must be considered within a definite family having the same typical construction and like components. However, these cost family curves are subject to much change from year to year due to the extremes in the cost-reduction curves that are prevalent with such material items as solid-state components, hybrids, and integrated circuits.

The nomogram, or formula, Fig. 3-2, reconciles all these families for method 2 application.

CONCLUSIONS AND MAJOR POINTS

1. Method 2 is recommended for estimating subsystems, functions within systems, units, or strictly measurable functions.
2. Constants can be developed for any type of item for which historical costs are available. The variances in the historical costs make these data useless as-is for estimating. Removing the variances from the data will provide a reliable constant to which the variances are then replaced according to the forecast sensitive elements. (See Fig. 3-8)
3. A project should be broken into secondary or lesser elements for greater visibility and accuracy if time permits. One or two of the other five parametric estimating methods should also be used in conjunction with method 2.

METHOD II, UNIT CONSTANTS (1970 Monetary Base)

Unit Description	Item "C" – Dollars
Computer (summary) (P. S. & Memory not incl.)	9.30
Comparator	5.02
Computer Units	8.04
Generator – Pulse	6.28
Memory	4.58
Controls & Displays (summary)	4.91
Indicators	6.13
Selectors	3.92
Miscellaneous	
Amplifiers	4.05
Converters	6.04
Flight Sensing	5.21
Racks (use system volume)	11.68
Receiver – I R	2.36
Relay Assembly	1.84
Selector – Channel	1.33
Test Set	4.58
Power Supply (summary)	1.38
Filters – D. C. power	3.37
Generators – AC & DC	0.45
Power Supplies	1.05
Reactors	0.22
Regulators	1.80
Terminal Boxes	1.25
Radar Antenna – (summary) *	223.64
Antenna Servo	14.18
Attenuator	9.69
Controller	4.75
Cooling *	95.63
Pressure Group *	58.27
Programmer	19.89
Radar Receiver – Transmitter (summary)	1.46
Receiver only (summary)	4.42
Parametric Amplifier	3.87
Transmitter – Radar (summary)	1.22
Transmitter – only	1.39
Oscillator – Radar	2.37
Power Supply – HV	3.27
Filters – Doppler	2.98
Oscillator – R. F.	4.16
Synchronizer	4.05
Switch Box	4.42
Waveguide *	303.20
Timer – Stop	7.47

* These items seldom have the volume given; thus barring the calculation of a density index. A density index of 1 (one) is used in these cases.

Fig. 3-8. Typical unit constants for method 2.

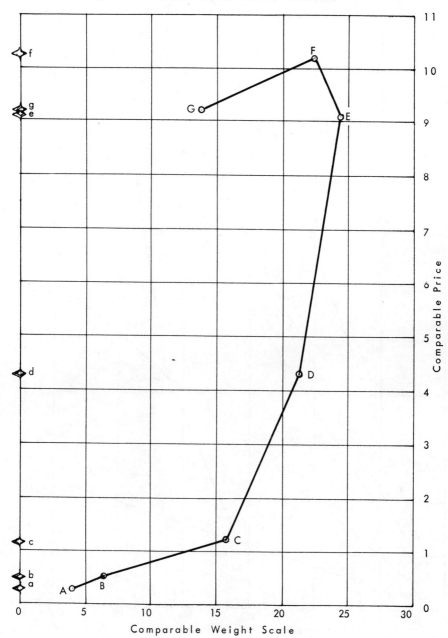

TRENDS: PRICE RELATIVE TO WEIGHT

Fig. 3-9. Trends: price relative to weight.

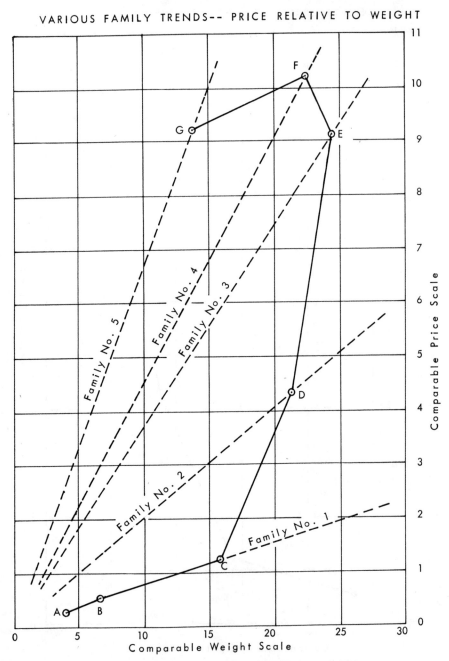

Fig. 3-10. Various family trends: price relative to weight.

4. The newness element is most critical, but one must try to weigh the possibilities. One of the major new components may suddenly become common due to its broad application and overproduction, resulting in the bottom dropping out of the price. This may happen twice in a row, but it is not likely that a pattern has been established. The next time it may be that the project you are estimating will be one of its few users.

5. Constants and curves for any type of engineering can be developed in the same manner as that used in developing the method of estimating the implementation costs.

4

Method 3: Parameters for Budgeting Men, Materials, and Money

I rediscovered this method in a hotel room as I was faced by six questioning engineers who had just finished roughly describing a far-out program. It was now my show and I had to come up with something fast, so I assumed that certain tasks would be required, and posted them down the left side of a large work sheet; then asked if I had missed any unique task.

The engineer who was most closely associated with a specific task was asked to suggest its starting date and completion; and was then encouraged to estimate the men required for each month. Material costs per month were also posted for some tasks as required.

The finished estimate was totaled and readily "dollarized" on an estimating form that displayed the total dollars with details of the dollars for each task.

The chief engineer then looked at me in amazement and said, "That is easy when it is done that way."

This is often the only method used for estimating many types of work.

Method 3 possibly should be part of most estimates even though the total costs of men and material have been estimated by other methods. The totals for any project should be spread by month and year, for this reveals the impact of a new program; it may indicate that not sufficient men are available, and that a certain portion must be subcontracted if the facilities are not expanded.

The text on method 3 deals with budgeting manpower and the total dollars only, because material requirements are unique to each program and their spread depends on the expedience perceived by the planner.

MANPOWER PARAMETERS USED IN ESTIMATING

This estimating method 3 is basically a spreading of the requirements for the respective tasks for a program over the period of performance to provide costs for a coordinated plan to complete the program on schedule. It is usually performed at the men and material level first; then, as soon as the labor and material have been extended to total cost, a spread of total commitments and expenditures can be made.

The task consists of planning the program within set parameters and making management decisions.

1. How soon can the project be started, and when must it be completed?
2. What are the required tasks, or types of work?
3. When can each task start and finish so that it may best relate to other tasks?
4. How many men per task are required each month?
5. What tasks require material, and how much per month?

Figure 4-1 illustrates a typical manpower chart.* It is based on yearly quarters instead of months for this illustration, but months are usually more practical. Listing the required tasks and their parameters requires professional assistance.

Best results are obtained by comparing the proposed project to historical data on similar projects, and making notes of any comparative estimates that were used; at some later date it may be well worth this added effort.

Most tasks require a buildup from a minimum starting level to a maximum level that may be for one month only, or for a long period after which there is a fast drop-off until zero requirements are the completion of the task.

---

*Called a Gantt chart, named for its originator.

Quarter	1	2-	3	4	1	2	3	4	1	2	3	4	1	2	3	4
Delivery Schedule: Preproduction									4	6						
Production											8	24	36	50	50	32
ENGINEERING																
Design --------	5	10	4													
Prod. Eng. -----		10	20	8												
Loft & Temp. ---		2	4	3												
Prototypes ------		10	20	20	8											
Test & Rwk. ----			6	12	5											
Total Eng.	5	32	54	43	13											
MANUFACTURING																
Planning ------				5	10	12	12	10	9	6						
Tool Design ----					4	8	5									
Tool Mfg. ------						8	16	10								
Test Eq. Eng. ---				5	10	12	8									
Test Eq. Mfg. ---					5	10	10	6								
Total Implementation				10	29	50	51	26	9	6						
PRE PROD. MFG.																
Fabrication ------						2	5	8	8	1						
Assembly ------							4	10	16	2						
Test ------								1	2	3						
Total Preproduction						2	10	20	27	6						
PRODUCTION MFG.																
Fabrication																
Sht. Mtl. ------									2	5	10	10	10	7		
Mach. Shop ------									1	3	6	7	7	4		
Processing ------									1	2	2	2	2	2		
Assembly																
Subassembly ------										4	12	18	20	20	16	
Test ------										1	3	6	7	7	6	
Assembly ------										1	12	25	25	25	18	
Test ------										1	2	3	3	3	3	
Syst. Bld.-Up & Test -										1	3	5	5	5	5	5
Total Production									4	18	50	76	79	73	48	5
GRAND TOTAL	5	32	54	53	42	52	61	46	40	30	50	76	79	73	48	5

* The beginning and end of a period of time.

Fig. 4-1. Method 3, manpower chart (Gantt chart).

The main purpose of this discussion of method 3 is to supply a couple of aids in spreading the estimated totals.

SPREADING MANPOWER, HOURS, OR DOLLARS – GRAPHIC METHOD

There are several estimating and management problems that can be solved if one is in possession of a fast, practical method of spreading the total costs, or manpower requirements. I prefer the method presented here in Figs. 4-2 and 4-3 because the spread is defined by the known parameters; it is fast, and I am prejudiced because of having developed it.

For example, let us assume a program of 25,000 hr to extend over a period of 15 months. Past records indicate a buildup of 5 months for this type of work and a 3-months drop-off period.

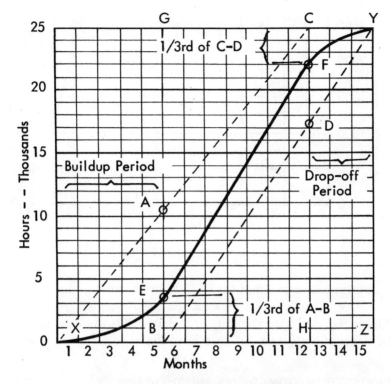

Fig. 4-2. Graphic distribution, cumulative.

The 15 months are indicated by equal spaces on the baseline X-Z, and the 25,000 hr are laid out on the vertical scale. The 5-month buildup is noted by a vertical line B-G at the end of the fifth month; and the beginning of the drop-off period is noted by line C-H.

The program will start at point X and follow some type of S curve to point Y; its course should be within the quadrilateral $XCYB$. Point E can be located at $\frac{1}{3}$ the distance A-B, and point F at $\frac{1}{3}$ the distance C-D. A straight line between E and F determines the input plateau. The spaces between the points X to E and F to Y can be filled in with the aid of a French curve.

Peak manpower requirements can be calculated by subtracting the total at E (3600 hr) from the total at F (22,500 hr) to equal 18,900

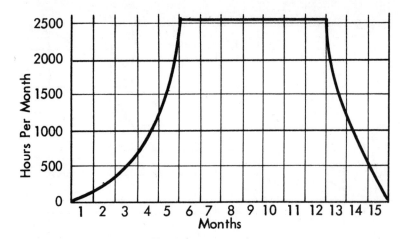

Fig. 4-3. Monthly plot of Fig. 4-2.

hr to be expended in 7 months at 2700 hr per month; and 167 hr per man per month indicates a peak manpower of 16 men.

Figure 4-2 provides the data for charts showing the monthly requirements in manpower or hours as in Fig. 4-3.

PERIOD DISTRIBUTION CURVES AND THEIR USE

My period distribution curve tables have been used extensively for manually spreading manpower and dollars; they have also been programed for computer application. They provide a fast method with six types of curves for a choice of a slow, fast, or average buildup, plus a plateau if desired. The tables are in Appendix 1; they are for the percent per period as this is the most useful form.

It is interesting to compare the results of using the curve tables on the same program data as used in the graphic method presented in Figs. 4-2 and 4-3. The table for curve 6 is the best fit, and the column designating 15 periods provides the monthly factors to apply to the 25,000 hr as detailed in the following table:

Month	Curve 6 (% per month)	Hours per Month	Cumulative Hours
1	0.5	125	125
2	1.4	350	475
3	3.4	850	1,325
4	5.9	1,475	2,800
5	7.8	1,950	4,750
6	8.9	2,225	6,975
7	9.7	2,425	9,400
8	9.7	2,425	11,825
9	9.8	2,450	14,275
10	9.7	2,425	16,700
11	9.7	2,425	19,125
12	9.3	2,325	21,450
13	8.2	2,050	23,500
14	4.8	1,200	24,700
15	1.2	300	25,000

Figures 4-4 and 4-5 graphically display these results with but minor variations from Fig. 4-2.

CONCLUSIONS AND MAJOR POINTS

1. Method 3 is one of the most widely used procedures for estimating and planning.
2. Unusual programs can be estimated in conference by this method.
3. Every estimate should have the benefit of the details derived from this method, plus its added confirmation.
4. All values incorporated in an estimate that are determined by estimating a programs relationship to historical data should be documented.
5. Spreading dollars by the graphic method is best for extreme cases, for example: A manager may wish to commit 80% of material purchases in 2 months after the "go-ahead" date and be at the 95% level at the end of 3 months; this is just a French curve exercise.
6. The period distribution curve tables in Appendix 1 have been extended through 25 periods for this book to easily accommodate up to a 2-year spread.

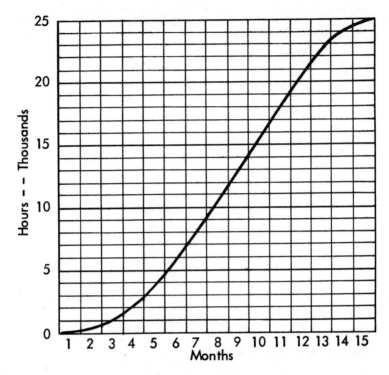

Fig. 4-4. Spread of hours by tables, cumulative.

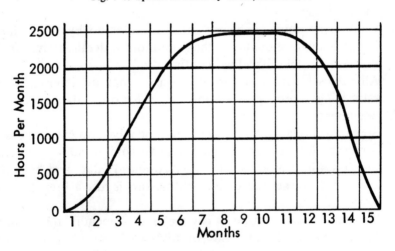

Fig. 4-5. Spread of hours by tables, monthly.

5

Method 4: Parameters by Type of Work and/or Material

The simplicity of this method makes it possible to develop estimating standards for any firm in possibly the shortest time of any method. Estimating standards (dollars per unit of output) can be developed and applied to each type of work to arrive at the estimated total cost for any project. These standards can be used for setting up an estimating procedure that will promote consistency by one or several estimators. Each standard can be established from records, or by inspection and observation.

These standards can be as broad or as detailed as is practical. For certain types of industry, a straight cost for each type of product might be sufficient, whereas other types of business may require hourly standards by each type of work, by department, or by cost center.

Method 4 standards must be for the same elements used in typical estimating data, those elements used in describing new projects. Certain types of woodworking plants have found that the standard for any department, or by cost center, could be based on the number of board feet processed in that area. This greatly simplifies estimating, for each type of item requires a certain number of board feet for which the costs are built up according to standards for the cost centers involved.

Other companies perform certain types of work on specific materials; this may require standards by type of material, by type of work. Due to the quantity and variations in these classifications only general suggestions can be given to aid in establishing the needed estimating standards.

DEVELOPING A STANDARD

A survey of the reports put out by each department will often suggest the unit of measure for many of the departments. For a sheet metal department it may be the square feet of material processed, the number of boxes produced, or the number of pounds of material used, etc. For the machine shop there are usually ways to develop a common unit of measure by cost center. This also applies to electrical and electronic standards. Painting or processing standards can be by the square foot of the respective finishes or processes.

It is in this area of arriving at a unit of measure that imagination and initiative come into play. Sometimes it seems impossible to measure certain types of work, but persistence often turns up the unexpected.

Example: Assume the following information for a sheet metal department. Reports are available giving the number of pounds of material used per month for several months. It is also possible to obtain the amount of production hours for the department for the same months (see Fig. 5-1). All that remains to be done is to find a way of tying the number of pounds produced to the labor hours, so that the information is useful as a tool.

If the pounds processed per month are plotted relative to the hours for each respective month as shown in Fig. 5-2, a line can be drawn through the points as shown by $X-Y$. For extreme accuracy this line should be located mathematically, but for some practical purposes it can be done by inspection. This is done by considering half the length of the line at a time and making the sum of the distances to the points on each side of the line equal.

In the absence of reports it may even be necessary to count the workers in a department and obtain the count of items produced from shop records or foreman's notes.

THE STANDARD AND POSSIBLE VARIATIONS

In Fig. 5-3, line $X-Y$ represents the hours per pound, but these hours per pound are not the same for a 2000-lb production load as they are for a 3000-lb load. For instance, 2125 hr are required for 2000 lb, or 1.1 hr/lb, and 3300 hr are required for 5000 lb, or 0.66 hr/lb. Therefore the anticipated load for the respective department for the duration of the program must be located on the chart in order to forecast the hours per pound for the standard.

Month	Production Hrs.	No. Lb Processed
(1) Jan.	1600	1000
(2) Feb.	2360	3725
(3) Mar.	3175	5400
(4) Apr.	3360	4000
(5) May	2400	2800
(6) June	2400	2000

Fig. 5-1. Production data.

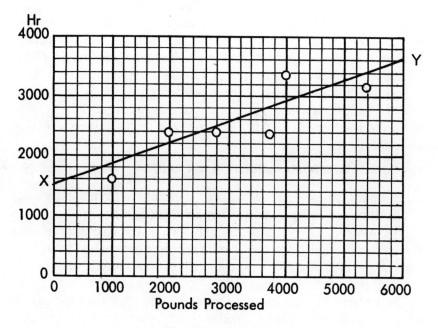

Fig. 5-2. Linear average.

To read the estimated hours directly from this chart, line *a–b* can be drawn through zero and the forecast load at *b*. By moving the decimal points of both scales the same number of digits, the proper range for direct reading can be made; thus, 1000 lb would require 700 hr; 10 lb would require 7.0 hr.

Armed with this type of chart for each class of work, it is possible to estimate a new project by breaking it down into various units of production and reading the required hours from the respective charts.

Figure 5-4 continues and expands the use of Fig. 5-3. On each side of the point (*b*) that was chosen as the forecast department load, previous periods have experienced loads higher toward *Y*, or lower toward *X*. This being the case, it is well to develop possible high and low standards also. This is done by selecting a typical medium light load point, *c*, and a higher load point at *d*. These will produce comparable estimates. Top management can then have greater visibility for choosing the estimate that is indicated by forecast contingencies.

Figure 5-4 is also a type of break-even chart; the shaded area to the left of *b* is loss, and that to the right of *b* is added profit, so any deviation of the forecast load to the left of *b* is very undesirable.

THE ESTIMATING FORM – AN IMPORTANT AID

An estimating form is needed to keep this method from being merely a collection of unrelated charts, thus eliminating redundancy of operation coverage. The form should supplement the charts, promoting ease of application. One or more lines can be used as needed for the input from each chart and to record the unit of measure that is used to develop the total for the item.

The form in Fig. 5-5 contains some of the important elements. It is given only to help present the idea clearly. If the particular requirements of a company are built into a form it can be invaluable.

It may be found that some elements will not require charts, merely a standard amount each that can be put directly on the form. This can then be used as a multiplier of the total value involved.

Various designations can be used in the "type of work" column: R for rough, A for average, P for precision, etc. These will indicate a percentage that must be added or deleted for work inconsistent with the average. The proper factor can then be circled to describe the work, and also to provide the adjusting amount.

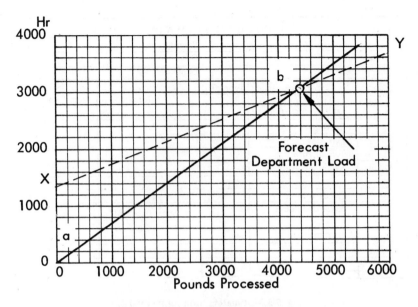

Fig. 5-3. Hours per pound chart for estimating.

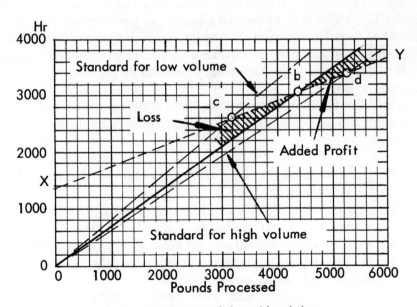

Fig. 5-4. Hours per pound chart with variations.

P R O J E C T   E S T I M A T E

Item No.				Program No.	
Description				Date	
Engineer				Estimator	

Chart No.	Description	Type of Work			Value in Total	Total Hours From Chart
		R	A	P		
	MECHANICAL					
1	Castings				Lb	
2	Extrusions				Ft	
3	Turned Parts				Qty	
4	Machined Parts				Qty	
	SHEET METAL					
5	Boxes and Racks				Sq Ft	
	ELECTRICAL					
6	Harnesses				Wires	
	Etc.					

Fig. 5-5. Estimating form for method 4.

This type of estimating incorporates all known items in a readily usable form, taking full advantage of all the available facts. It is often easier to tie this type of estimating into the performance of a company, or department, than may be imagined. For an estimator starting with a new company it provides a very good interim method, and it is often preferable to depending on standards and related data developed at some other firm.

The following routine can be used in developing a feasibility model for method 4 and to outline the general requirements for an estimating form.

1. List the various types of work with their subelements down the left side of the form.
2. Retrieve all applicable charts on hand. (One may be amazed at the number that can be used for an initial model.)
3. Rough-out charts for items not covered by charts on hand.
4. Try out the model on several programs to determine any weak points or additional elements required.
5. Check tentative charts with current reports and revise as required.

IMPLEMENTATION – METHOD 4 ESTIMATING TECHNIQUES

Basic Data and Strategy

Experienced implementation costs for current and historical programs can be used to develop reliable estimating standards that can be applied with confidence in estimating the dollars required to implement planned programs.

It is first necessary to adjust all data so that there are no variances in the totals due to inflation or unusual conditions that may prevail at different time periods. These data can then be used in several ways. They can be used in total, by type of work, by cost center, etc., for their use is dependent only on the categories, or level, for which they were developed.

Implementation is relative to two elements; first, the amount of each type of work per product unit, measured in pounds, square feet, linear feet, or any other unit of measure that seems most appropriate for a rough evaluation of the output of a particular production group. The second sensitive element affecting implementation costs is the maximum planned rate of production per month.

If these two elements are known for a number of programs, a formula can be developed that will provide implementation estimates comparable to experienced costs on similar programs.

The Nomographic Method

If a computer is available, some cost engineers may find it ideal for developing any needed formulas; but simple nomagraphic techniques can first be used to advantage to analyze the problem, and also provide very satisfactory results in a form that can produce immediate results.

Comprehending the characteristics of the data is the first step. This is sometimes achieved by investigating various possibilities by trial alignment charts. After several attempts, one is usually sufficiently familiar with the data to choose one course and pursue it until a chart of best fit for the data is formulated.

The typical data in Fig. 5-6 are used for a case illustration in the development of an applicable nomogram and formula. These data are composed of ten historical programs. In actual practice it may

Project Description	Pounds of Product Z	Rate Per Month	Implementation Total Cost
No. 1	25	10	117,500
No. 2	125	6	464,000
No. 3	30	15	169,000
No. 4	40	8	170,000
No. 5	200	30	1,538,000
No. 6	150	15	846,000
No. 7	15	100	196,000
No. 8	75	7	298,000
No. 9	90	60	942,000
No. 10	140	70	1,560,000

Fig. 5-6. Given data on implementation costs.

sometimes be necessary to start with just a couple of programs and revise the formula as soon as any new data are available.

These data seem rather consistent, and they are possibly sufficient to develop a reliable formula for any new program within the range covered. The implementation cost element of the data is complicated, for no two programs are similar due to different amounts of product and different production rates; but if we divide each implementation cost figure by the pounds of product of the respective project, we can develop an implementation cost table for one pound of product for each of the given rates as in Fig. 5-7. The production rates and the implementation cost per pound of product Z from this table can then be plotted on arithmetic grid graph paper as in Fig. 5-8. A line drawn by inspection through these data points provides the comparable implementation costs per pound of product Z for all production rates within the range of the given data.

We could stop here, assuming that the problem was solved, for all that is needed to estimate a new program is to locate the planned rate of production value on the rate scale of Fig. 5-8 and note where its vertical line intersects the line A–B; this point when followed on its horizontal line to the left-hand scale locates the point that indicates the implementation dollars for one pound of product Z at the planned

Project Number	(1) Pounds of Product Z	(2) Rate Per Month	(3) Implementation Cost	(3) - (1) = (4) Implementation Per Lb. Prod Z	(5) Implementation Chart Estimate
No. 1	25	10	117,500	4,790	120,000
No. 2	125	6	464,000	3,712	470,000
No. 3	30	15	169,000	5,635	170,000
No. 4	40	8	170,000	4,250	170,000
No. 5	200	30	1,538,000	7,690	1,500,000
No. 6	150	15	846,000	5,640	850,000
No. 7	15	100	196,000	13,276	200,000
No. 8	75	7	298,000	3,972	300,000
No. 9	90	60	942,000	10,467	950,000
No. 10	140	70	1,560,000	11,143	1,600,000
			6,300,500		6,330,000

Fig. 5-7. Given data with implementation cost per pound at respective rates and comparable estimates read from the nomogram.

rate. Multiplying this implementation cost per pound index by the unit pounds of product Z within the new project develops the estimated total implementation dollars for the new project. However, we can simplify the task by developing a simple nomogram that supplies the full implementation cost for product Z by one alignment operation based on the total pounds of product Z and the maximum rate of production planned for the new program.

The arithmetic grid paper as first used for plotting the pounds and rate data results in a curve that is difficult to define (see Fig. 5-8); but log-log paper is convenient for plotting data of this nature because it often develops a curve that can be approximated by a straight line on log-log paper. Figure 5-9 is the result of using log-log paper for plotting these data, and it provides the needed straight line. It is very simple to develop a multiplication nomogram using the data from this log-log scale chart because one type of multiplication nomogram uses logarithmic scales, making use of the same general principle as used for a slide rule.

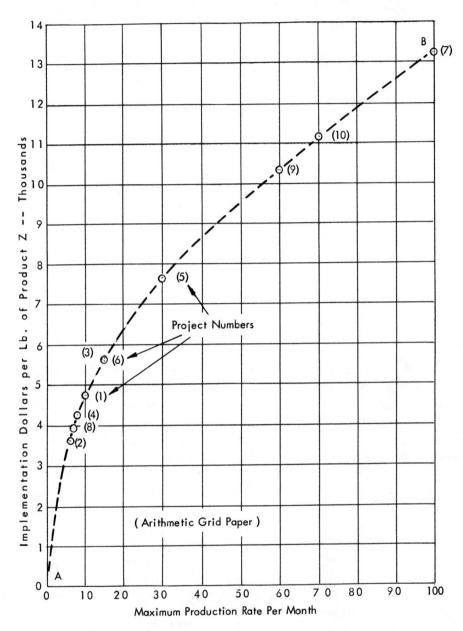

Fig. 5-8. Implementation dollars per pound of product Z versus the respective production rates per month.

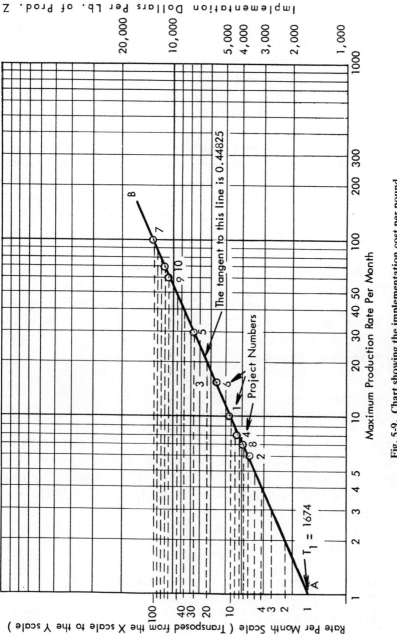

Fig. 5-9. Chart showing the implementation cost per pound
for all rates within the range of the data.

The logarithmic scale of the graph paper, Fig. 5-9, can be used for the nomogram, Fig. 5-10, in the following manner: Two and one-half cycles of the logarithmic scale of the chart, Fig. 5-9, are used for the 200-lb range required for the "unit pounds of product Z" scale. The scale* for the rate per month is the "implementation dollars per pound" scale in Fig. 5-9 with the various rate of production values projected on it, and the rates noted thereon. This scale with the projected rates is transferred to the nomogram for the right-hand scale. Provisions have now been made for the two outside scales of the nomogram. The center scale of the nomogram is placed halfway between these two scales; and the logarithmic scale for it is composed of cycles one-half the length of the cycles in the "pounds of product Z" scale. Proper numerical values can be placed on this center scale by merely applying some of the given data and thus determining the number of significant figures.

The nomogram can now be checked by reading from the center scale the implementation dollars for each project of the given data. Variations of the implementation dollars provided by the chart can be related to the data dollars for each project and for the sum of all projects; all can be noted as a check for accuracy. Figure 5-7 makes this comparison in column 5.

THE ESTIMATING OF PRODUCTION SUPPORT, OR ALLOCABLE COSTS AND SERVICES

Production support is a summary of the costs of materials, labor, and related applied expenses incurred in the support provided by various production engineering and other organizations to the groups engaged in the actual manufacturing of the product. It may include the following typical types of support:

1. Maintenance of parts lists
2. Change processing
3. Planning updating to reflect manufacturing changes
4. Tool repair and maintenance
5. Trouble-shooting of test equipment

---

*Space limitations required the cycles of Fig. 5-9 to be less than those of Fig. 5-10.

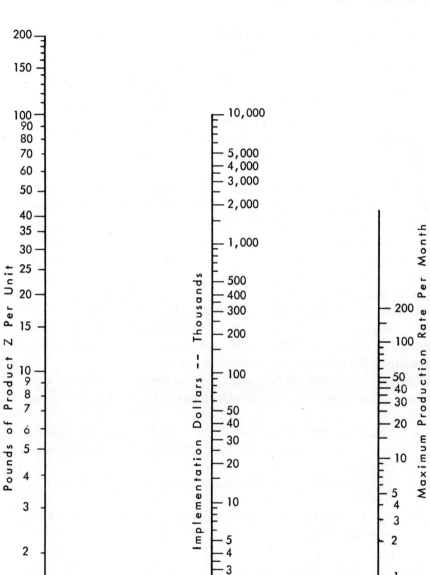

Fig. 5-10. Nomographic solution to the formula for estimating product Z implementation.

6. Test equipment engineering in support of existing equipment
7. Maintenance of existing equipment
8. Material expended during the performance of the above tasks
9. Quality engineering in maintaining existing procedures
10. Laboratory effort expended in maintaining or recalibration of instruments and equipment
11. Program management during the production phase
12. Production coordination (production control)

These costs may be estimated by the responsible departments in various manners, depending on the project and the type of support being considered. A summary of these estimates, or actuals, from the various departments provides a total dollar value for the production support. These totals from representative programs can be compared by percentage relationship to the respective basic factory cost of each, resulting in percentages that are not of a fixed amount, but are variable within a certain range. An investigation of the extremes shows that a high percent of support is needed on a small initial production quantity as compared to the much smaller percent required for a large program with a high rate of production per month.

The sensitive elements causing this range in percent of production support versus basic manufacturing cost can be divided thus:

1. There is a certain amount of labor and material necessary to provide services that are largely relative to the manufacturing time span.
2. There are other costs that relate more closely with the production rate.
3. The type of product has a certain control over each of the first two elements.

Considering these sensitive elements, it is possible to develop formulas based on previous detailed estimates or historical records of completed programs. This can be done in total as presented in Fig. 5-13. The same procedure can be used to develop the formulas for estimating each type of support effort, thus providing for the unique mix of support effort for various types of products.

A simple method of developing needed formulas of this type is presented to illustrate the principles involved.

Developing a Formula Using a Graphic Technique

1. Lay out parallel, vertical scales as in Fig. 5-11; one scale for Contract Span in Months, and the other for the Production Rate per Month.

2. Through the proper Rate per Month and Contract Span points on these scales draw connecting straight lines for each of the representative production support estimates that are being used for determining the formula. Place notes on each of these lines recording the percent of basic manufacturing cost that is equal to the respective support amounts.

3. These lines, by their points of intersection, and percent value, can be used to determine the location and calibration of the center scale, representing the percent of basic manufacturing cost.

4. Calibrating the center scale to correct inconsistent points and to determine the intermediate values is a very simple task if approached in the following manner:

   Reproduce the center scale with the projected points as determined by the representative programs; this scale to be the Y axis of a graph as in Fig. 5-12, and a X axis having sufficient equal divisions to represent the full desired range of the "percent of manufacturing cost" scale. The Y scale can then be given the proper numbers.

   All of the projected points that were transferred from the center scale of Fig. 5-11 can be used to determine a curve on Fig. 5-12 by plotting each point's equiavalent location relative to the Y scale.

   A curve of best fit is drawn through these points which can be used in reference to the X scale to determine the complete calibration of the Y scale, which is then transferred to Fig. 5-11 and the finished nomogram, Fig. 5-13.

This chart can then be reduced to a formula if it is desirable to use a strictly mathematical formula for estimating future production support tasks, but it is much easier to use the developed nomogram, Fig. 5-13, and a straightedge for obtaining an estimate that is sufficiently precise. It should be satisfactory for most cases simply to read the percentage from the nomogram and to apply this factor to the estimated basic manufacturing cost.

PRODUCTION SUPPORT

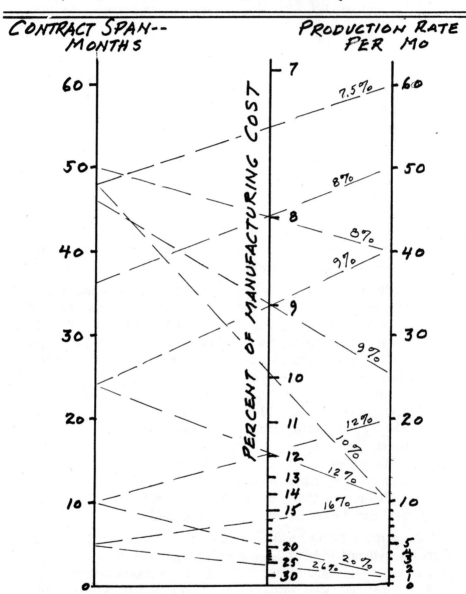

Fig. 5-11. Developing the production support formula.

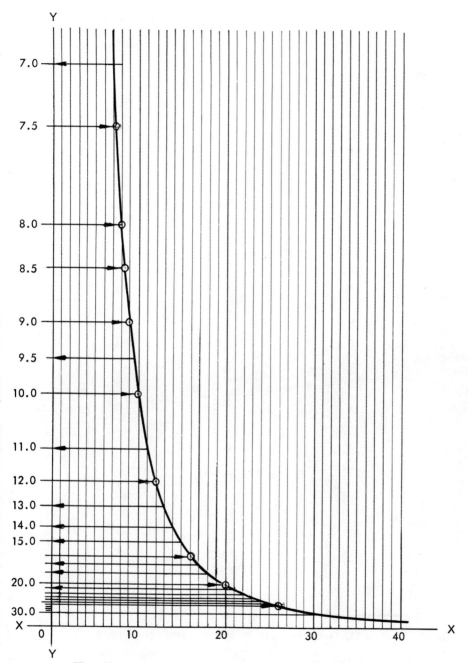

Fig. 5-12. Developing the calibration of a scale resulting from a curve.

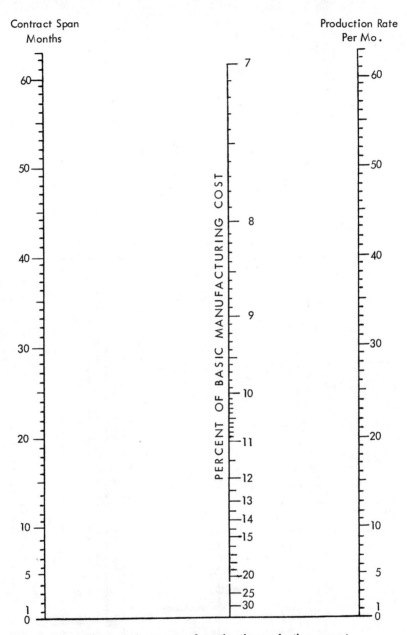

Fig. 5-13. Typical nomogram for estimating production support.

Instructions for Use of the Nomogram

Sensitive Elements Defined

Contract Span, Months
The period of time that production support is charged; roughly, the number of months in which delivery is made.

Production Rate per Month
The average rate—the total quantity divided by the contract span in months.

Example

Consider a 24-month program span with an average delivery rate of 15 items per month.

Method: Locate the points indicating the 24-month contract span and 15 per month rate on the respective scales of Fig. 5-13. Lay a straightedge through these two points; then, the intersection of the straightedge with the center scale of the nomogram should locate the point indicating the estimated percent to apply to the basic manufacturing cost for this example. The estimate as per the nomogram is 11.2%.

CONCLUSIONS AND MAJOR POINTS

1. Method 4 provides a ready means of meshing estimates with a factory's performance.
2. Most fabrication and assembly hours, or costs, can be developed as illustrated, but it may be necessary to develop some hypothetical formulas for certain assembly and quality control items not directly covered.
3. The simplest feasibility model may prove to be superior to any method previously available.
4. Warning: There may be a problem when estimating a new program that is for fewer items than represented by the typical program used to develop the data for the standards. For instance, if the data are for an average of 300 items each, and the proposed program is for 100 items, it is necessary to use some simple learning-curve techniques to make the proper adjustment.

5. The estimating form can save much time and reduce the possibility of errors.
6. It pays to rough-out a feasibility model, for it gives direction to the operations required for the final version.
7. Revise all tentative charts and standards to conform with contemporary records and performance.
8. The implementation nomogram gives one phase of nomogram construction, and the study on production support shows one method of using nomograms for regression analysis.

6

Method 5: Parameters for Modular Work Measurement

I believe that this is the best method of estimating that is available to date, because it is applicable to any stage of engineering. It provides reliable estimates plus other data that are usually not available until a program is half complete.

It can possibly be assumed that the best method for estimating an item is to retrieve all records on a very similar item and substitute the identification of the new part for that of the previously produced item. This is usually impossible, but method 5, a statistical system, can provide the equivalent so that the final estimate is as though one had a detailed planning sheet for each part, with each operation timed. The system can also include the estimated assembly, quality control, etc., plus the implementation.

Method 5 is very simple and logical, using the essential production planning and estimating know-how of specialists for each type of work without enlisting their active participation; the data banks of their applied estimates supply these basic data used in developing method 5 nomograms and formulas. It is recommended that these basic planning data estimates be by some form of work measurement system such as MTM (Methods Time Measurement) to assure uniformity.

The initial MTM system dealt with extremely minute time and motion values (micro data). These data have been summarized into combinations at much higher levels with little loss of accuracy to reach MTM-5 (macro data), which in combinations are used for setting time standards (standard hours) for planned operations on parts and assemblies. These operations standards are combined statistically to form the modular data used in method 5. This puts the final link in a chain comprised of setting standards for performance, performance reporting, data banks, and estimating standards.

DEVELOPING MODULAR DATA

Method 5 eliminates the wasted labor and mass of paper work involved in making detailed estimates for parts and assemblies, etc. The results by method 5 are more applicable and consistent, their having a base of standard hours and plant performance of the respective plant.

Selecting a particular type of item that varies in size, and then charting the relative cost, or production time, versus size is common practice; it is a fine method for estimating parts. This results in a trend for estimators to collect piles of such charts, which are useless for estimating complete assemblies, for they lack adaptability and usually have needless and redundant operations when used in various combinations. Method 5 contains the key for developing formulas for modular data so that all elements work together.

Method 5 modular data are developed by selecting typical parts for all needed categories, with full planning and standards (MTM or other) applied to each operation; then through simple analysis, charts and formulas are developed for applying standard hours (in total) to fabrication, assembly, and quality control operations. All data must:

1. Include the labor to fabricate and process the work. Any fabrication labor to provide for installing other parts must be deleted.
2. Include fabrication operations necessary for installation into the next higher part, or assembly.
3. Include installation, plus handling time.

This is the key to method 5's compatible standards for modules, insuring full coverage without redundancy.

An estimating form will unify all formulas and charts into a coordinated system. Figure 6-1 presents some of the general requirements by type of labor or task.

Method V Modular Plan by Work Categories	Fabricate Part	Fab. Mating Hole, Recess, or Area	Retrieve Part, Orient and Install	Weld, Braze, Etc.	Quality Control
Mechanical					
Castings --------	X		X		X
Gears ----------	X	X	X		X
Machined Parts --	X	X	X		X
Bolts, Etc. ------		X	X		X
Other ----------					
Sheet Metal					
Sheet Mtl. Pts. --	X		X		X
Screws, Rivets, Etc.		X	X		X
Welding --------				X	X
Brazing --------				X	X
Electrical					
Printed Ckt. Bds.	X		X		X
Harnesses -------			X		X
Plugs -----------		X	X		X
Tubes -----------		X	X		X
Transistors ------			X		X
Diodes ----------			X		X
Multi-term. Comp.			X		X
Pig-tail Comp. --			X		X
Other ----------					

Fig. 6-1. Typical labor categories per type of item.

Method 5 History and Formulas for Computer Programing

Method 5 was developed with the following goals in mind. It must be:

1. A method that would not soon become outdated; it must be adaptable to new techniques and grow with changes in the state of the art.
2. Planned to provide consistent results at all times.
3. A method that would reflect the capability of the firm using it.
4. Simple and accurate, easy to use.
5. Able to allow changes in an estimate to be made readily.
6. Geared to making three or more estimates to reflect probabilities without extensive back-up data changes.
7. Equal to supplying documented back-up data for each estimate.
8. A method whereby the engineers could communicate their best thinking at the time, with an indication of the anticipated state of the art to be used.

Nomograms were used as the final analysis step in developing formulas from experience data. They were used because they are a very good regression analysis technique that can be used to prove the theory involved before translating them into formulas. Formulas for all method 5 nomograms are given in Figs. 6-10 through 6-14.

All method 5 nomograms are included because a full grasp of the strategy and operations of method 5 is necessary before computer assist can be realized to the fullest. See Appendix 3 for method 5 nomograms and a typical estimate.

Standard hours were used as a base because this opens up all planning done on work in production as one great data bank; and one can draw on the knowledge and expertise of specialists in their respective fields. Only this approach can provide the needed intelligent estimates because this is a most practical way of covering all phases of production and making estimates consistent with the respective plant capacity.

Have the goals been met? The changes and additions over the 20 years since the initial method 5 program have been insignificant; never has it become outdated. However, some of the component categories were more important then than now. The results are consistent because the base is calculated and all items of judgment are noted.

Any item estimate that is done in total for which there is a delivery schedule, general description, and component count can be fully dollarized for implementation and production costs in approximately one hour. A complex unit, fully engineered, can be processed by method 5 in a couple hours. This does not include the tabulating and pricing of parts and material from the engineering drawings, which may require from one to five days.

Changes can be readily made by changing the hours-per-standard-hour estimate only, without disturbing the base standard-hours estimate. Additional estimates can be easily made by variations in the hours-per-standard-hour factor and forecast material prices without disturbing the base.

Method 5 provides a complete record of the details and assumptions of the estimate.

Method 5 was developed during the early days of computers when all available computer time was exclusively assigned to problems other than estimating. In the meantime, the initial nomograms provided a satisfactory means of applying method 5 formulas. Since then, the state of the art and the availability of computers have opened the door for a completely new era in estimating, and method 5 provides the first step in the right direction. Method 5 formulas are provided in this book, along with all other details of method 5 so that a computer program can be developed. It is recommended that this be done first in an abbreviated form, and then additions can be intelligently made, thus promoting the full potential of method 5.

Mechanical Modular Data Standards

Keeping this system within workable bounds requires a few compromises. It is best to start with a minimum number of categories and formulas because method 5 can grow as new categories are required.

Gears present such a case: It was found that the gears could be grouped, and the needed formulas reduced to only three, based on diameter and grade of precision.

For castings and machined parts the weight and quantity of the rough stock are the only values readily available for making conceptual estimates; these values are also applicable for estimating fully engineered items. (See Fig. 6-2.)

The plotting of the standards for various sizes of castings by type of material revealed that the standard-hours-per-pound formulas for aluminum and steel are so nearly identical that the same formula could be applied to either. The same is true for machined parts, the reason being that steel versus aluminum are related three to one by both weight and machinability, thus reducing the formulas involved by 50%.

Setup time should be part of the module standards and be based on the experience of the respective firm. That is, the cumulative quantity at which the cumulative average of the actual hours is equal to that of the standard hours will require one setup per x items; then the setup time divided by the quantity per setup is the standard. Setup time will then vary, being subject to the hours-per-standard-hour factor.

To make the module complete it is necessary to include the fabrication time for drilling, facing, relieving, etc., that are required in the next higher receiving part.

Assembly of each component into the next higher item is part of the modular standard concept. It is therefore necessary to obtain the standard time for installing and handling the part through the full course of assembly.

Sheet Metal Modular Data Standards

The modular concept for sheet metal is based, first, on fabrication, processing, and/or painting of the part completely, exclusive of method of assembly. Thus, it is necessary simply to add the riveting time for a riveted assembly, the welding time (spot or arc) for a welded assembly, etc., and the assembly time, which is the handling time through the assembly operations. (See Fig. 6-3.)

The basic modular standard for sheet metal fabrication time is well represented by the square feet or weight of the required sheet metal per part, by type of material. Firms using aluminum, steel, and stainless steel may find it preferable and reasonably accurate to use pounds per part and thus may avoid having a formula for each type of material.

MODULAR CONCEPT FOR GEARS OR MACHINED PARTS

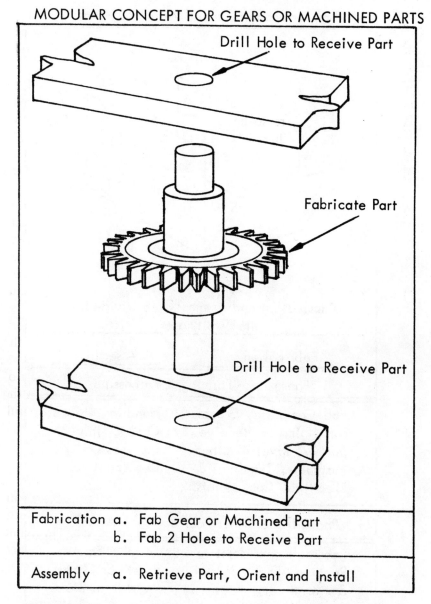

Drill Hole to Receive Part

Fabricate Part

Drill Hole to Receive Part

Fabrication a. Fab Gear or Machined Part
b. Fab 2 Holes to Receive Part

Assembly   a. Retrieve Part, Orient and Install

Fig. 6-2. Typical mechanical module: modular concept for gears or machined parts.

Plan for Modular Sheet Metal Standards

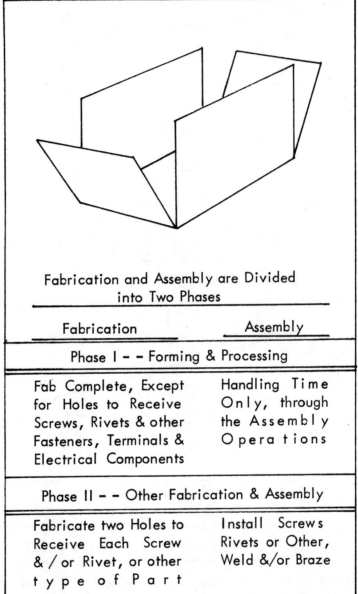

Fabrication and Assembly are Divided
into Two Phases

Fabrication	Assembly
Phase I - - Forming & Processing	
Fab Complete, Except for Holes to Receive Screws, Rivets & other Fasteners, Terminals & Electrical Components	Handling Time Only, through the Assembly Operations
Phase II - - Other Fabrication & Assembly	
Fabricate two Holes to Receive Each Screw & / or Rivet, or other type of Part	Install Screws Rivets or Other, Weld &/or Braze

Fig. 6-3. The sheet metal module: plan for modular sheet metal standards.

Calculating the weight of the material for metal parts is very simple if one observes that the weight of 0.063 aluminum is 0.0063 lb in.^{2} ; the decimal point on the thickness figure when moved one place to the left equals the weight of aluminum per square inch. For steel or brass one need but multiply the weight of aluminum by three. Magnesium weighs two-thirds as much as aluminum.

In developing the modular standard from planning sheets with standard time per operation, it is necessary to remove any standards for punching or drilling holes to receive rivets or other parts. For each rivet or screw, the time to fabricate two holes is included, plus the assembly time to install each rivet or screw. Spot or arc welding can be added relative to the number of welds and the total length of the weld.

Electronic Modular Data Standards

Printed Circuit Boards. There are several types of single-layer boards, so a formula is required for each. Multiple-layer boards are made by stacking boards of the required circuits; hence, multiple-layer boards are two or more single-layer boards.

Harnesses. The two obvious variations in harnesses are the average length of wires and complexity; these can be used to develop the basic module standard. The attaching of any plugs is included in that the soldering-in of both ends of each wire is included.

Electronic Components. The basic module standard can assume hand soldering of each terminal. This is true of small quantities only, since wave soldering and other methods are used on large quantities and this results in reduced time which is covered by the hours-per-standard-hour factor developed from the use of learning-curve techniques to obtain an adjusting factor.

Plugs and Connectors. A hole is often punched for each plug or connector and two to four screws can be added for each, the electrical connections being covered in the soldering-in of the harnesses.

Tubes. Miniature tubes can be considered in the same category as multiterminal components. Other tubes require the fabrication and assembly required to mount their sockets on chassis or printed circuit boards; plus the retrieval, orientation, and pressing in place of the tube.

Transistors and Multiterminal Components. Heavy items such as transformers are usually bolted down, so each should call for four bolts, the electrical being usually covered by the harness soldering-in. Transistors and lightweight multiple-terminal components are often mounted by soldering-in on chassis or printed circuit boards. Retrieval, orientation, and soldering of each terminal should be included.

Estimating the Hours-Per-Standard-Hour Factor

The use of standard hours for the basic estimate requires a forecast of the hours-per-standard-hour factor in order to complete the estimate. This is done by the application of simple learning-curve techniques.

Learning-Curve Determinants. Learning curves have two controlling features: the value at unit one and the slope of the curve. These two elements provide the key to evaluating experience data and developing formulas for applying the results to proposed programs in the proper measure to reflect forecast conditions, for example, type of program, quantity, rate of production.

Certain elements determine the hours per standard hour at unit one and others govern the slope.

Elements Determining the Hours-Per-Standard-Hour at Unit One. The phase of program, whether it be prototypes, preproduction, or production items, establishes the general category area for the hours per standard hour at unit one. Then, newness of the type of program and the complexity of the product are the key elements. These can all be included in Method 5 formulas.

Elements Governing the Learning-curve Slope. The hours-per-standard-hour value at unit one is an important element in determining the slope of a learning curve, because a completely new program will start with an extra high hours-per-standard-hour figure and hence a steep learning curve should be realized, whereas the typical program will start at a low point for unit one and have little opportunity for learning. The program's rate of production is also of major importance. Even a casual analysis of the results of minimum and maximum production rates will illustrate this point.

Developing the Formula for Hours Per Standard Hour at Unit One

A decision must be made concerning the level at which the hours-per-standard-hour factor is to be applied. It would be ideal to develop this factor for each element of work, or by cost center in all categories. This can overburden the system with details, so it may be best to set up the initial model with an hours-per-standard-hour factor for basic divisions only.

It is well to place the experience data in the three categories of: prototypes, preproduction, and production; this will provide for the class of work. Therefore three charts or formulas will be required. The two elements remaining are newness and complexity. Sometimes the newness scale will provide sufficient definition for all three classes of work, one formula therefore being sufficient as in Fig. 6-4.

Much can be accomplished in performing the needed analyses by selecting the best examples, keying them into some type of nomogram that will aid in making a specific analysis, and then determining the basic formula to make all other data consistent within reasonable limits.

A simple but positive scale is needed for both newness and complexity (see Fig. 6-4) in order to analyze historical data by nomography and to develop an hours-per-standard-hour formula. These scales can vary by type of work if definition is improved. The laying out of the data relative to the newness and complexity scales will quickly suggest the center scale of best fit. Figure 6-4 is a typical hours-per-standard-hour-at-unit-one nomogram for sheet metal work.

An applicable formula can then be written to replace the nomogram if it is preferable to chart reading, or if it is needed for computer application.

Developing the Formula for the Learning-curve Percent

The rate of production is one of the major sensitive elements of learning-curve slopes. Much of the available learning-curve literature has been based on data developed in war years when rates of production were high and men were frozen to their jobs. As a result, little evidence of the results of low production schedules was possible. This has caused many firms to quote too low, based on anticipated curves that did not reflect the contract's reduced production rate.

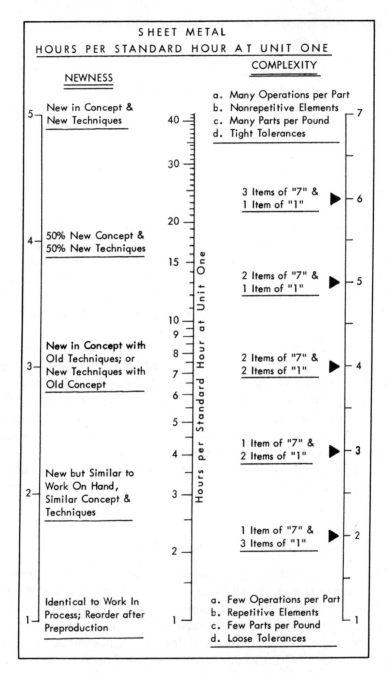

Fig. 6-4. Formula for hours per standard hour at T_1.

A minimal production rate results in simple tools and the assignment of men to each station, with little chance of ever reducing their number. This makes for a very flat learning curve.

The schedule that starts low and increases to a maximum can well show improvement, for the tooling can be constantly improved and the number of operations required per man can be reduced, which usually results in ample improvement.

Hours per standard hour at unit one as previously determined from Fig. 6-4 provides the other major determinant. A new type of program will start high with much learning possible, but familiar types of programs start low, and little improvement can be gained.

A typical nomogram was made for sheet metal (see Fig. 6-5), with a center scale covering the working range of learning-curve slopes. The historical data indicated that the slope did not drop as fast as the rate increased, so the rate scale was made logarithmic for a trial; this indicated that the hours per standard hour at unit one was also logarithmic. All the data were then plotted and the scales adjusted for best fit for all data. The nomogram was then translated into a formula to be used for computer programing.

Hours Per Standard Hour for the Program

Once the learning curve has been established by determination of the hours per standard hour at unit one and the slope percentage, the task of calculating the hours-per-standard-hour factor for the program is a simple learning-curve technique.

Figure 6-6 provides a graphic approximation method. Unit learning curves are used so that both initial programs and follow-on programs can be processed by the same chart.

Step 1. The midpoint of the program is the first requirement. For an initial program, a straightedge is laid through the first unit at *A* and the last unit of the program on line *B–C*; then the point at which it crosses line *T–U* determines the approximate unit of average value. An example for a follow-on program is given as part of Fig. 6-6. A straightedge is laid through the first unit (*H*) as located on line *D–E* and the last unit (*I*) as located on line *F–G*, indicating a midpoint at *J*.

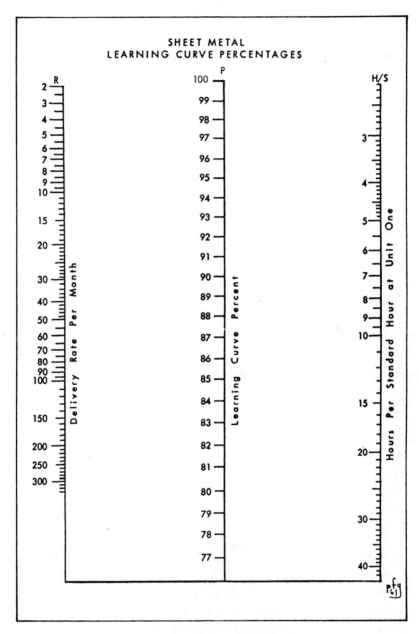

Fig. 6-5. Sheet metal learning-curve percentage nomogram.

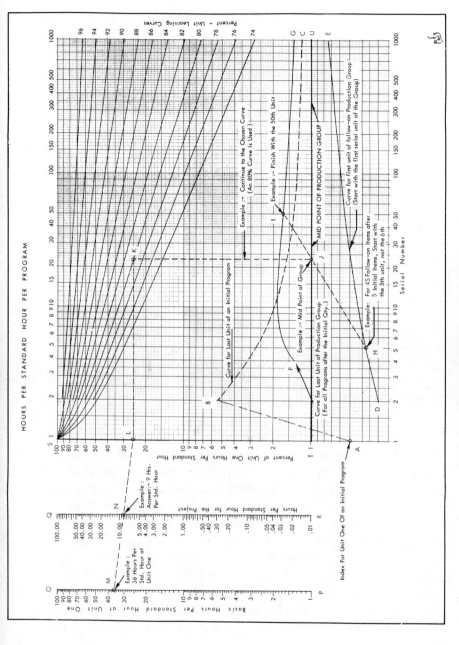

Fig. 6-6. Method 5. Hours-per-standard-hour nomogram.

Step 2. The midpoint on the selected learning curve is located directly above *J* at point *K* on the required curve.

Step 3. The point *K* determines the point *L*, which indicates that the hours per standard hour for the program are 25% of the hours per standard hour at unit one; this multiplication is done in step 4.

Step 4. The hours per standard hour at unit one are located on line *O–P* at *M* and a straightedge is laid through *L–M*; this determines the hours per standard hour for the program at point *N*. This is shown as 9 hr.

The calculating of the hours per standard hour for a program can also be computed by using the following formula with X_a as the first unit of the program and X_b as the last unit, Q as the program quantity, and the value of L from the schedule, Fig. 10-4.

$$T_P = \left\{ \left[X_b^L - (X_a - 1)^L \right] \div Q \right\} \cdot T_1$$

This formula can be resolved into an easily used pocket calculator routine.

METHOD 5 FORMS

The Work Sheet

The first of the two forms required for method 5 is a work sheet similar to that shown in Fig. 6-7. It is laid out so that it can be used to summarize a rough list of components for a conceptual estimate or for an indentured bill of material for a fully engineered item. The center portion of the sheet is composed of three sections: one for mechanical items, the next for sheet metal, and the last for electrical items. These three sections provide for listing of the values of the respective sensitive elements used in calculating the modular standards.

The right-hand side of the sheet provides space for tabulating the amounts and prices of the various materials and parts used. Provisions are made for three categories: purchased parts, subcontract, and other, plus the totals column.

METHOD NO. 5 WORK SHEET   Date ....   Ref No. ....   Sheet / of /

Project: EXAMPLE

Unit No. 987654-100

Unit Name BLACK-BOX

Indentured Item Description (1 2 3 4 5 6)	Qty. Req'd	Castings Q	Castings #	Gears Q	Gears Θ	Machined Parts Q	Machined Parts #	Bolts, Screws Q	Boxes & Racks P	Boxes & Racks □'	Screws Q	Rivets Q	Terminals Q	Welding - Heliarc P	Welding - Heliarc In.	Brazing P	Brazing Cu. In.	Spot-Welding P	Spot-Welding S
MECHANICAL																			
Castings	10	10	20																
Gears	24			24	1.5														
Machined Pts.	75					75	8												
Bolts	24							24											
Screws	50							50											
SHEET METAL																			
Sht. Mtl.	60								60	15									
Screws	90										90								
Rivets	240											240							
Terminals	150												150						
Welding														10	20			16	150
ELECTRICAL																			
Pr. Cir. Bds.	40																		
Harnesses	1																		
Connectors	6																		
Plugs	40																		
Indicator Lights	10																		
Transistors	80																		
Diodes	322																		
Transformers	8																		
Coils	10																		
Capacitors	250																		
Resistors	300																		
Subtotal											90	240	150						
Factor %											100	60	30						
Totals		10	20	24	1.5	75	8	74	60	15	90	144	45	10	20			16	150

Indentured Item Description	Circuit Wldg. Q	Printed Circuit Bds. Q	Printed Circuit Bds. □"	Harnesses / Cables W	Harnesses / Cables In.	Plugs Q	Tubes W.S. Q	Miniature Tubes Q	Miniature Tubes T	Transistors Q	Diodes Q	Multi-Term. Components Q	Multi-Term. Components T	Pig-Tail Comp. Q	Wire Wrap Q	Wire Wrap □"
Pr. Cir. Bds.		40	30													
Harnesses				450	20											
Connectors						6										
Plugs						40										
Indicator Lights							10									
Transistors										80						
Diodes											322					
Transformers												8	6			
Coils												10	4			
Capacitors														250		
Resistors														300		
Totals		40	30	450	20	46	10			80	322	18	5	550		

Indentured Item Description	Kind or Description	Size	Unit Cost	Purchased Parts	Sub-Contract	Other	Total
Castings	Aluminum		#5.00/#		#100.00		#100.00
Gears	Steel		0.30			#7.20	7.20
Machined Pts.	Steel		1.00/#			8.00	8.00
Bolts			0.15	#3.60			3.60
Screws			0.10	5.00			5.00
Sht. Mtl.	Aluminum		#1.00/#			15.00	15.00
Screws			0.08	7.20			7.20
Rivets				0.80			0.80
Terminals			0.10	15.00			15.00
Pr. Cir. Bds.			0.90			36.00	36.00
Harnesses			0.08			60.00	60.00
Connectors			5.00	30.00			30.00
Plugs			3.00	120.00			120.00
Indicator Lights			3.00	30.00			30.00
Transistors	Silicon		6.00	480.00			480.00
Diodes			0.50	161.00			161.00
Transformers			20.00	160.00			160.00
Coils			4.00	40.00			40.00
Capacitors			1.00	250.00			250.00
Resistors			0.40	120.00			120.00
Totals				1422.60	100.00	126.20	1648.80

Q = Quantity
# = Lbs. of Raw Mat.
Θ = Average Diam.

P = Number of Parts
□' = Square Ft. of Raw Mat.
S = Quantity of Spots
In. = Inches of Weld
Cu. In. = Cubic In. Part Size

□" = Square Inches - Average Size
W = Quantity of Wires
T = Number of Terminals

Estimator ....
Remarks: -

Fig. 6-7. Method 5. Work sheet for summarizing data.

After all of the details have been listed, whether it takes one or a dozen pages, they are summarized into one line at the bottom of the last sheet, for only the totals are needed for method 5. These totals are then transposed to the estimating form as shown in Fig. 6-8.

Details, such as those available for a typical black box in the conceptual stage, are included and summarized in Fig. 6-7. Note that this form documents the anticipated requirements by types of parts and their quantities as of the date of the estimate. Subsequent changes are expected, and usually made, but these changes can be readily noted and their affect on the total price can be weighed and documented. This makes for good relations and promotes the credibility of the estimating.

Some firms may need only one of the three categories, such as sheet metal, and others may find it advisable to have a separate sheet for each type of work, but the results can be the same.

The Estimating Form

An estimating form is controlled by its usage. The form in Fig. 6-8 was designed to include all the pertinent data for the complete product estimate. It even provides for alternate quantities, or for two estimates such as minimum and target.

The values in total of the sensitive elements from the work sheet are posted in their respective places under the three categories: mechanical, sheet metal, and electrical; these are listed vertically on the estimating form rather than horizontally as on the work sheet.

The next step is to determine the standard hours by applying the modular data formulas or charts. These are posted in the columns for standard hours under fabrication, assembly, and test. Then the hours per-standard-hour factors for the various types of work are calculated by formulas or charts that were developed for this purpose.

Standard hours are then summarized by type of work and multiplied by the estimated hours-per-standard-hour factors to provide the hourly totals as illustrated by the posted example, Fig. 6-8.

Several estimates can be completed at the bottom of the sheet by extending the hours to dollars and including the various other elements plus the material costs as estimated on the work sheet. The example is for alternate quantities of 46 and 100.

Unit No. 987654-100 Name BLACK - BOX

Size 10 x 12 x 20 Similar to: (No. .... (Name ....

Wt. 30 Lb

Engineer .... Phone .... Estimator ....

LABOR ESTIMATE: 46 Ea. @ 5/Mo. PROGRAM — Start Date .... End Date .... From Unit No. 5 Through Unit No. 50.

Alternative Qty. 100 Ea. @ 5/Mo. PROGRAM — Start Date .... End Date .... From Unit No. 5 Through Unit No. 104.

Reference No. .... Date: ....

CHART NUMBER ITEM	Min.	Av.	Max.	VALUE 1 Am't. - Unit	VALUE 2 Am't. - Unit	STANDARD HR Fab.	STANDARD HR Ass'y	STANDARD HR Test	46 Ea. HR/STD. H. F.	A.	T.	PROGRAM ESTIMATE FAB.	ASS'Y	TEST	TOTAL HOURS	Alt. 100 Ea. HR/STD. H. F.	A.	T.	PROGRAM ESTIMATE FAB.	ASS'Y	TEST	TOTAL HOURS	MATERIAL ESTIMATE COST EACH	TOTAL COST	
MECHANICAL:																									
1. Castings				10 Parts	20 Lb	17.3	1.8					($T_1$8.5	8.2	8.1)					($T_1$8.5	8.2	8.1)				
2. Gears		✓		24 Parts	1.5" Av Θ	18.5	5.0																		
3. Machined Parts		✓		75 Parts	8 Lb	18.9	9.5					(C%88	90.3	96)					(C%88	90.3	96)				
4. Bolts; Screws; Etc.				74 Parts	10 Cstg.*Q.	3.2	1.2																		
5.																									
Total - Mechanical						57.9	17.5		3.9	4.4		225.8	77.0		302.8	3.4	3.8		196.9	66.5		263.4			
6. Inspect & Test Assemblies								2.1			6.2			13.0	13.0			5.9			12.4	12.4			
SHEET METAL:																									
7. Boxes; Racks; Chassis; Etc.				60 Parts	15 Sq. Ft.	4.0	3.4																		
8. Screws; Rivets; Terminals				279 Parts	60 S.M. Pts	2.4	5.4					($T_1$11.2	11.2	8.1)					($T_1$11.2	11.2	8.1)				
9. Welding - Heliarc Hand				10 Parts	20 In.		0.3																		
10. Brazing; Silver Soldering				Parts	Cu. In.							(C%91.5	91.5	96)					(C%91.5	91.5	96)				
11. Spot Welding				16 Parts	150 Welds		0.5																		
12.																									
Total - Sheet Metal						6.4	9.6		6.3	6.3		40.3	60.5		100.8	5.8	5.8		37.1	55.7		92.8			
13. Inspect & Test Assemblies								0.7			6.2			4.3	4.3			5.9			4.1	4.1			
ELECTRICAL:																									
14. Circuit Welding				Welds	--																				
15. Printed-Circuit Boards			2s	40 Qty.	30 **	36.0																			
16. Harnesses	✓			450 Wires	20 Length		28.5																		
17. Plugs - All Types				46 Qty.		0.6	2.4																		
18. Tubes with Sockets				10 Qty.	5 Q.Chass	0.2	0.2																		
19. Miniature Tubes				Qty.	A. Term.																				
20. Transistors				80 Qty.	--		2.3																		
21. Diodes				322 Qty.	--		5.9					($T_1$11.2	28.2	11.8)					($T_1$11.2	28.2	11.8)				
22. Multiterm. Components				18 Qty.	5 A. Term.		1.0																		
23. Pig-Tail Components				550 Qty.	--		11.9					(C%91.5	82.6	93.7)					(C%91.5	82.6	93.7)				
24. Wire Wrap				Bds.	**																				
25.																									
Total - Electrical						36.8	52.2		6.3	8.9		231.8	464.6		696.4	5.8	6.8		213.4	355.0		568.4			
26. Inspect & Test Assemblies								7.8			7.6			59.3	59.3			7.0			54.6	54.6			
SUBTOTAL						101.1	79.3	10.6																	
27. System Test								8.7			7.6	($T_1$11.8	C%93.7)	66.1	66.1			7.0	($T_1$11.8	C%93.7)	60.9	60.9			
GRAND TOTAL						101.1	79.3	19.3				497.9	602.1	142.7	1,242.7				447.4	477.2	132.0	1056.6		1648.80	1731.00

	FABRICATION Hours	Rate	Labor $	Oh.@150	ASSEMBLY Hours	Rate	Labor $	Oh.@100	Prem. @1%	Subtotal	Material	M.B. @10%	Subtotal	Other	Subtotal	Prod. Supp.	Total Cost	Pkg.	Factory Cost	G&A6% F/P.10%	Approx. Price
Each For 46	498	3.00	1494	2241	745	2.00	1490	1490	30	6745	1,731	173	8649			150	8799	50	8849		10,422
Total For 46	22,908				34,270						79,626									(Factor)	479,412
Each For 100	447	3.00	1341	2012	609	2.00	1218	1218	26	5815	1,731	173	7719			140	7859	40	7899	1.1778	9,303
Total For 100	44,700				60,900						173,100										930,300

Note: No's. 5, 12, and 25 have been left free for unusual elements or growth.

* Or Machined Parts Receiving Bolts, Screws, etc.

** Average Board Size - Square Inches

Mechanical Standard Hours: Total 128.2

Electrical Standard Hours: Total 71.5

Fig. 6-8. Method 5. Estimating form with completed estimate.

An abbreviated form may be desirable for computer input usage.

This estimating form consolidates all formulas or charts into a unified system that is based on all the data available for a product at the time of the estimate, no matter what the stage of the engineering.

ESTIMATING THE REQUIRED
IMPLEMENTATION OR ENGINEERING

Standard hours are a definite index to the complexity of a product, and the purpose of engineering and implementation are to produce a specific product. It is therefore logical to relate engineering and implementation to the total product standard hours (see Fig. 6-9).

Some of the facets of implementation are:

> Production engineering and planning.
> Tool planning, design, and engineering.
> Test equipment planning, engineering, and design.
> Tool and test equipment manufacturing.
> Purchased tools and test equipment.

These can all be estimated by relating them to the program's total standard hours.

The historical data that were at hand for developing the initial implementation formulas for method 5 were for typical programs that varied from 90% mechanical to 85% electronic. This suggested the need for dividing the total standard hours into two divisions, the mechanical portion, and the electronic portion plus quality control and test. Hence, the total mechanical standard hours and the total electronic standard hours provided two scales for a nomogram on which the historical data could be plotted after it had been adjusted for a specific production rate per month by use of a chart similar to Fig. 3-6. Thus it was possible to estimate a particular type of engineering, or implementation, for a basic rate per month and then extrapolate to the required rate for the program being considered. The results have always proved to be amazingly reliable and consistent.

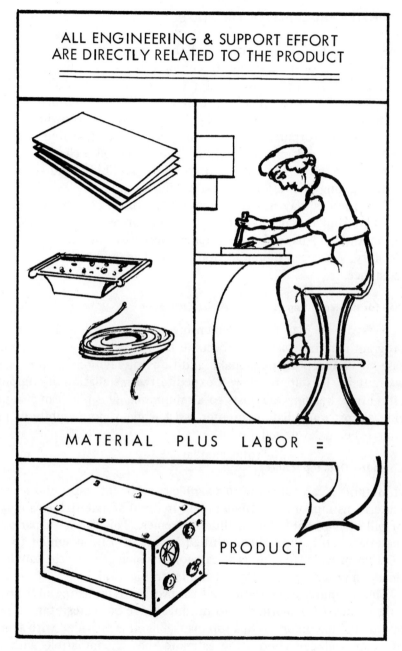

Fig. 6-9. Standard hours. An index of the complexity of a product.

COMPUTER PROGRAM FOR METHOD 5

The availability of computers and the benefits that can be realized from their use opens the way for computer assisted estimating.

Method 5 is an ideal system for taking advantage of present computer capability with minimum cost and effort. An outline for computer programing of method 5 is given in Fig. 6-10, showing how all elements and formulas function to produce the estimated costs for both product and implementation. These method 5 charts and their formulas are given as an illustration of the feasibility and simplicity of programing this very logical estimating method.

All formulas may not fit the precise methods used by any particular firm, but this method may give superior results to any estimating method on hand for consistency, accuracy, amount of program visibility, and speed of application, even when used without computer assistance.

Plan for Implementing Computer Assist Estimating

A favorable plan for implementing computerized estimating is to program method 5 as given; become familiar with its advantages and ease of application; then make additions or revisions as they seem warranted. This approach will expedite the installation of computer estimating, keeping the costs to a minimum. This plan can also provide almost immediate computer assist while a more ambitious program is being developed.

Computer Data Input Sheet

Method 5 can function on a simple component count plus a rough material listing, or a complete indentured bill of material, or any type of bill of material between these extremes. This listing of parts and materials can be best accomplished by use of a form similar to Fig. 6-7 that provides for the basic categories which can then be tabulated on a computer input data sheet similar to Fig. 6-11.

The computer input data sheet has provisions for listing all the basic data required for method 5 to produce three estimates: target, minimum, and maximum. Thus top management is provided with a range of costs and can choose the estimate that is compatible with the market and business risks.

METHOD V OUTLINE FOR COMPUTER PROGRAMMING

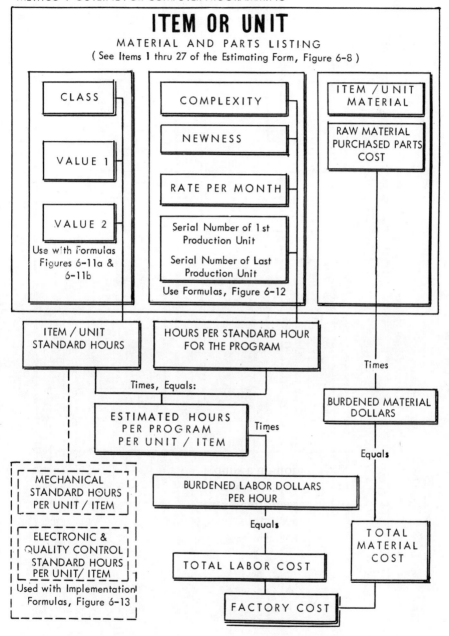

Fig. 6-10. Method 5 outline for computer programing.

COMPUTER INPUT DATA SHEET

Unit No........... Name ....   Ref. No. .... Date ....

PROGRAM: Start Date...... End Date .... Rate Per Month ....   Engineer .... Phone ....

Type of Program: Preproduction... Initial Prod. Qty. ...Direct Follow-on .... Restart Follow-on Prod. Qty. ....Interim ....

TABLE NUMBER ITEM	CLASS Min	CLASS Ave	CLASS Max	VALUE 1 Amt.–Unit	VALUE 2 Amt.–Unit		COST PROBABILITY RANGE — Min. Estimate Fab.	Assy.	Test	Target Estimate Fab.	Assy.	Test	Max. Estimate Fab.	Assy.	Test
MECHANICAL															
1. Castings				Parts	Lbs.	Newness									
2. Gears				Parts	Ave. O	Rating									
3. Machined Parts				Parts	Lbs.	Complex									
4. Bolts; Screws; Etc.				Qty.	Q. Cstg.	Rating									
SHEET METAL															
7. Boxes; Racks; Chassis; Etc				Parts	Sq. Ft.	Newness									
8. Screws;				Parts	S.M.Pts.	Rating									
Rivets				Parts	S.M.Pts.	Complex									
Terminals				Parts	S.M.Pts.	Rating									
9. Welding – Heliarc				Parts	Inches										
10. Brazing; Silver Soldering				Parts	Inches										
11. Spot Welding				Parts	Welds										
ELECTRICAL															
14. Circuit Welding				Welds	-----										
15. Printed Circuit Boards				Qty.	A.S.I.*	Newness									
16. Harnesses				Wires	Length	Rating									
17. Plugs – All Types				Qty.	-----	Complex									
18. Tubes With Sockets				Qty.	Q.Chas.	Rating									
19. Miniature Tubes				Qty.	A.Term.										
20. Transistors				Qty.	-----										
21. Diodes				Qty.	-----										
22. Multi-Term. Components				Qty.	A.Term.										
23. Pig-Tail Components				Qty.	-----										

(See Appendix III, Figures AIII-32 thru AIII-39 for guidance in selecting Newness and Complexity Indices)

* A.S.I. - - Average Square Inches

Fig. 6-11.

Standard-Hours Formulas

All of the needed formulas for standard hours are given in Figs. 6-12a and 6-12b. These formulas are for the respective items listed on the computer input data sheet, Fig. 6-11. Familiarity with the nomograms and their respective coverage notations in Appendix 3 is necessary to fully comprehend each formula. These formulas can provide the basic standard hours for an item or unit.

Learning-curve Formulas

It is necessary to forecast the learning-curve slope and the hours per standard hour at unit one prior to arriving at the hours per standard hour for the program. These formulas, (Fig. 6-13), use newness (forecast state of the art) and complexity indexes. These are easily and consistently estimated by selecting the needed indexes from the respective nomograms, Figs. A3-31 through A3-39. The learning-curve formulas provide the learning-curve percent and the hours per standard hour at unit one for use with the next formula to calculate the hours per standard hour for the program.

The Hours-Per-Standard-Hour-Per-Program Quantity

This formula, as expressed in the nomogram, Fig. 6-6, is translated into a formula as given in the last item of Fig. 6-13.

These data, formulas, and computer program outline should make the programing of method 5 as simple as possible because so much of the difficult work has been done and presented here. See Appendix 3 for additional information and an example of how method 5 is used to estimate a black box.

Implementation Estimating—Method 5 Computer Program Plan

The standard hours per item, or unit, of the production program should be keyed in such a manner as to summarize the electrical plus quality control standard hours into one subtotal and the mechanical standard hours into another subtotal. These two figures and the rate of production per month provide the data needed in the implementa-

MECHANICAL

	FABRICATION FORMULAS	ASSEMBLY FORMULAS
1. CASTINGS	$S_F = 0.32\, C_T + 0.71\, W_T$	$S_A = 0.032\, C_T + 0.071\, W_T$
2. GEARS — Comm.	$S_F = \text{Antilog}\left\{\left[\text{Log}\, G_T + 0.9\,(\text{Log}\, D_a)^{1.036}\right] - 0.33\right\}$	$S_A = 0.13\, G_T$
Prec. 1	$S_F = \text{Antilog}\left\{\left[\text{Log}\, G_T + 0.9\,(\text{Log}\, D_a)^{1.07}\right] - 0.26\right\}$	$S_A = 0.215\, G_T$
Prec. 2	$S_F = \text{Antilog}\left\{\left[\text{Log}\, G_T + 0.91\,(\text{Log}\, D_a)^{1.037}\right] - 0.0773\right\}$	$S_A = 0.33\, G_T$
3. MACH. PARTS Comm.	$S_F = 0.803\, W_T + 0.0913\, Q_P$	$S_A = 0.4015\, W_T + 0.0457\, Q_P$
Ave.	$S_F = 0.8\, W_T + 0.168\, Q_P$	$S_A = 0.4\, W_T + 0.084\, Q_P$
Prec.	$S_F = 0.82\, W_T + 0.22\, Q_P$	$S_A = 0.41\, W_T + 0.11\, Q_P$
4. BOLTS; SCREWS; ETC.	$S_F = 0.03405\, Q + 0.069\, C_T$	$S_A = 0.013\, Q + 0.02\, C_T$
5. (Spare for added item.)		
6. INSPECT & TEST		$S = 0.028\, S_M$

SHEET METAL

	FABRICATION FORMULAS	ASSEMBLY FORMULAS
7. BOXES; RACKS; CHASSIS; Etc.	$S_F = 0.035\, Q + 0.134\, A_G$	$S_A = 0.04\, Q + 0.0725\, A_G$
8. SCREWS	$S_F = 0.004\, Q + 0.02\, Q_P$	$S_A = 0.015\, Q + 0.02\, Q_P$
RIVETS	$S_F = 0.0024\, Q + 0.02\, Q_P$	$S_A = 0.009\, Q + 0.02\, Q_P$
TERMINALS	$S_F = 0.0012\, Q + 0.02\, Q_P$	$S_A = 0.0045\, Q + 0.02\, Q_P$
9. WELDING - HELIARC BY HAND		$S_H = 0.0076\, W_I + 0.0157\, Q_P$
TACK WELD		$S_T = 0.00393\, W_S + 0.016\, Q_P$
AUTOMATIC WELD		$S_A = 0.00275\, W_T + 0.016\, Q_P$
10. BRAZE; or S. SOLDER		$S_B = \text{Antilog}\left[(\text{Log}\, Q_A + 0.7103\, \text{Log}\, V) - 2\right]$
11. SPOT WELD Under 500 Cu. In.		$S_W = 0.008\, Q_P + 0.00215\, Q_W$
Over 500 Cu. In.		$S_W = 0.011\, Q_P + 0.00285\, Q_W$
12. (Spare for added item.)		
13. INSPECT & TEST		$S_{I+T} = 0.076\, S_A$

Fig. 6-12a.

	FABRICATION FORMULAS	ASSEMBLY FORMULAS

14. CIRCUIT WELDING — — — — — — — — — — — — — — — $S_A = 0.01\ Q_W$

15. PRINTED CIRCUIT BOARDS
Eyelet Type – One Side $S_F = \text{Antilog} \left\{ \left[\text{Log } Q + 0.858\ (\text{Log } A_a)^{1.083} \right] - 2.062 \right\}$
Eyelet Type – Two Side $S_F = \text{Antilog} \left\{ \left[\text{Log } Q + 0.7124\ (\text{Log } A_a)^{1.218} \right] - 1.765 \right\}$
Plated Thru — — — — — $S_F = \text{Antilog}\ (\ \text{Log } Q + 0.8572\ \text{Log } A_a - 1.28\ )$

16. HARNESSES — Simple — — — — — — — — — — — — — $S_T = \text{Antilog}\ (\ \text{Log } Q + 0.191\ \text{Log } L - 1.495\ )$
Average — — — — — — — — — — — — — $S_T = \text{Antilog}\ (\ \text{Log } Q + 0.239\ \text{Log } L - 1.509\ )$
Complex — — — — — — — — — — — $S_T = \text{Antilog}\ (\ \text{Log } Q + 0.318\ \text{Log } L - 1.55\ )$

17. PLUGS — — 1 per Pnl. — — — $S_F = 0.0226\ Q$ — — — — — $S_A = 0.0526\ Q$
2 Per Panel — — — — — $S_F = 0.0132\ Q$ — — — — — $S_A = 0.0526\ Q$

18. TUBES WITH SOCKETS $S_F = 0.0063\ Q + 0.02\ Q_C$ $S_A = 0.011\ Q + 0.02\ Q_C$

19. MINIATURE TUBES
With 3 Leads — — — — — — — — — — — — — — — $S_A = 0.0201\ Q + 0.0343\ Q_C$
With 4 Leads — — — — — — — — — — — — — — — $S_A = 0.0268\ Q + 0.0343\ Q_C$
With 5 Leads — — — — — — — — — — — — — — — $S_A = 0.0335\ Q + 0.0343\ Q_C$
With 6 Leads — — — — — — — — — — — — — — — $S_A = 0.0402\ Q + 0.0343\ Q_C$
With 7 Leads — — — — — — — — — — — — — — — $S_A = 0.0469\ Q + 0.0343\ Q_C$
With 8 Leads — — — — — — — — — — — — — — — $S_A = 0.0536\ Q + 0.0343\ Q_C$

20. TRANSISTORS — — — — — — — — — — — — — — — $S_A = 0.0304\ Q_C$
Xstrs. with Sleeves — — — — — — — — — — — — $S_A = 0.0359\ Q_C$
Xstrs. with Brackets — — — — — — — — — — — — $S_A = 0.0658\ Q_C$
Xstrs with Slvs. & Brkts. — — — — — — — — — — — $S_A = 0.0703\ Q_C$

21. DIODES — — — — — — — — — — — — — — — — — $S_A = 0.0193\ Q_C$

22. MULTI-TERM. COMPNTS.
On Plated Thru Bds. — — — — — — — — — — — — $S_A = \text{Antilog} \left[\text{Log } Q_C + 0.889\ (\text{Log } Q_T)^{1.074} - 1.886 \right]$
On Other Boards — — — — — — — — — — — — — $S_A = \text{Antilog} \left[\text{Log } Q_C + 0.504\ (\text{Log } Q_T)^{1.444} - 1.913 \right]$
Compnts. with Lugs — — — — — — — — — — — $S_A = 0.1\ Q_C$

23. PIG-TAIL COMPONENTS
On Eyelet Type Bds. — — — — — — — — — — — $S_A = 0.015\ Q$
On Plated Thru Bds. — — — — — — — — — — — $S_A = 0.0223\ Q$

24. & 25. (Spares for added items

26. ELECTRICAL ASSEMBLIES
Inspect and Test
Simple Units — — — — — — — — — — — — — — — $S_I = 0.06\ S_A$

Typical Units — — — — — — — — — — — — — — $S_I = 0.09\ S_A$
Complex Units — — — — — — — — — — — — — $S_I = 0.19\ S_A$

27. SYSTEM TEST
Ave. Electronic Eq. — — — — — — — — — — — $S_S = 0.046\ S_T$

Computers — — — — — — — — — — — — — — — $S_S = 0.08\ S_T$
Complex Equipment — — — — — — — — — — — — $S_S = 0.14\ S_T$
Unique Precision Equip. — — — — — — — — — — $S_S = 0.24\ S_T$

Fig. 6-12b.

LEARNING-CURVE FORMULAS

The following formulas are used to calculate the hours per standard hour per program for the respective types of work using the estimated standard hours as a base and adjusting them to the level of experienced actuals on previous programs.

LEARNING-CURVE FORMULAS (N = Newness & C = Complexity)
MECHANICAL FABRICATION & ASSEMBLY
Fabrication
Hrs / Std Hr at T_1: $\quad T_1 = \text{Antilog} \left[0.10102 \, (1.4975 \, N + C) + 0.12684 \right]$
Learning-Curve %: $\quad P = 93.4 - (1.72 \, \text{Log} \, R + 0.5 \, T_1)$
Assembly
Hrs / Std Hr at T_1: $\quad T_1 = \text{Antilog} \left[0.1124 \, (1.4975 \, N + C) + 0.02048 \right]$
Learning-Curve %: $\quad P = 100.1 - (5.3476 \, \text{Log} \, R + 6.7513 \, \text{Log} \, T_1)$
SHEET METAL FABRICATION & ASSEMBLY (Same formulas for both fab. & assem.)
Hrs / Std Hr at T_1: $\quad T_1 = \text{Antilog} \, (0.19 \, N + 0.1274 \, C - 0.2237)$
Learning-Curve %: $\quad P = 104.27 - (4.52 \, \text{Log} \, R + 9.2 \, \text{Log} \, T_1)$
ELECTRICAL FABRICATION & ASSEMBLY (Fabrication is sheet metal fabrication.)
Assembly
Hrs / Std Hr at T_1: $\quad T_1 = \text{Antilog} \left[0.709 + 0.00618 \, (N + C) \right]$
Learning-Curve %: $\quad P = 93.8 - \left[6.24 \, (\text{Log} \, R)^{0.815} + 4.455 \, (\text{Log} \, T_1)^{1.463} \right]$
ALL INSPECTION & TEST
Hrs / Std Hr at T_1: $\quad T_1 = \text{Antilog} \left[0.0053 \, (N + C) + 0.4344 \right]$
Learning-Curve %: $\quad P = 113 - (7.7 \, \text{Log} \, R + 12.9 \, \text{Log} \, T_1)$

The value of T_1 (Hrs / Std Hr at Unit One) and the learning-curve Percent as calculated by the above formulas are used in the following formula to determine the hours per standard hour for the various types of work for a program.

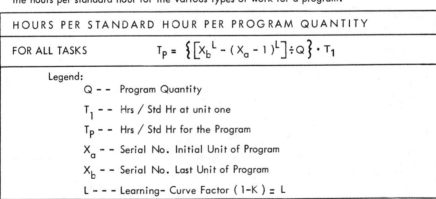

HOURS PER STANDARD HOUR PER PROGRAM QUANTITY
FOR ALL TASKS $\qquad T_P = \left\{ \left[X_b^L - (X_a - 1)^L \right] \div Q \right\} \cdot T_1$
Legend:
Q – – Program Quantity
T_1 – – Hrs / Std Hr at unit one
T_P – – Hrs / Std Hr for the Program
X_a – – Serial No. Initial Unit of Program
X_b – – Serial No. Last Unit of Program
L – – – Learning- Curve Factor (1-K) = L

Fig. 6-13.

tion formulas as given in Fig. 6-14. Additional information and nomograms for solving these forumlas are given in Appendix 3 with a typical implementation estimate for the black box used as an example.

CONCLUSIONS AND MAJOR POINTS

1. Method 5 is the needed link to form a complete chain of estimating by modular standards, the application of standards, and performance reporting. The reports are then built into the estimating formulas, and all previous planning and experience data become a supporting data bank.
2. The knowledge of specialists is incorporated without enlisting their direct participation.
3. Method 5 has a functional range that covers the initial product concept through the fully engineered version of the product.
4. It provides for making optional estimates for various quantities and conditions without changing the basic data. Modifications can therefore be readily made at management's request.
5. Estimating time is reduced to a minimum and yet it is in accord with the plant performance.
6. All fabrication and assembly are estimated in one pass for any mix of mechanical, sheet metal, and/or electronic products.
7. Method 5 provides firm, systematic documentation of the base used, and offers a means of comparative evaluation through the evolution of the engineering.
8. There are provisions for additions and growth.
9. Method 5 is presented as a complete system with charts, forms, and instructions in Appendix 3. This includes a typical conceptual implementation and production estimate for a black box with all readings noted on the nomograms used.

METHOD V- - IMPLEMENTATION FORMULAS

Basic Indices: E.   Electronic & Quality Control Standard Hours
                        System Tolal

                   M.   Mechanical Standard Hours, System Total

Legend:
    EB.   Engineering Base   S.   Standard Hours
    MB.   Mechanical Base   TE. Test Engineering
    P.   Planning   TD. Tool Design
    R.   Rate of Prod. or Del.   TEM. Test Equipment Mfg.
                                     TM. Tool Manufacturing

Charts I A & I B   Mechanical Base - - Tool Design - - 100 / Mo. Rate.
    Result used with Charts II & III.

$$MB \doteq 64\ M + 6E$$

Chart II   Tool Design for Program Production Rate

$$TD = \text{Antilog} \left[(\text{Log MB} + 0.1765\ \text{Log R}) - 0.4233 \right]$$

Chart III   Tool Manufacturing for Production Rate

$$TM = \text{Antilog} \left[(\text{Log MB} + 0.1765\ \text{Log R}) + 0.0835 \right]$$

Charts IV A & B   Engineering Base - - Test Engrg. for 100/ Mo. Rate.
    Result used with Charts V, VI, & VII.

$$EB = 2.1619\ M^{1.435} + 4.2344\ E^{1.435}$$

Chart V   Test Engineering for Program Production Rate

$$TE = \text{Antilog} \left[(\text{Log EB} + 0.285\ \text{Log R}) - 0.582 \right]$$

Chart VI   Test Equipment Mf. for Production Rate

$$TEM = \text{Antilog} \left[(\text{Log EB} + 0.3584\ R) - 0.7445 \right]$$

Chart VII   Planning for Program Production Rate

$$P = \text{Antilog} \left[(\text{Log EB} + 0.1263\ \text{Log R}) - 0.721 \right]$$

Tool Material, Purchased Tools, & Test Equipment Material are estimated using
dollars per hour factors for the respective items.

Fig. 6-14. Method 5 implementation formulas.

7
Final Evaluation and Probability Techniques

GENERAL

Estimating is forecasting the cost of a new product based on costs experienced on previous work that was similar. However, there is always the possibility that the cost may be higher or lower depending on a number of variable elements. This is true of an estimate in total and also true of each item of an estimate. Thus many items of cost, each with its own peculiar variable limits, are summarized in the total estimate. Hence, the costs of each item could vary so that they can be combined in many combinations. That is, if all of the actuals develop on the high side, the total cost will result in an overrun. The reverse can be true if the actuals are all on the low side, and there are all the other possibilities in between. These conditions are experienced in varying degrees, and exact performance on any item relative to the estimate is seldom a reality.

The labor estimate is very often based on the current productivity index which may not be applicable by the time the program gets into work. A sudden increase in the total factory load will necessitate the use of less experienced help, thus increasing the chance of an overrun.

Material estimates (other than firm quotes) are equally unstable. This is particularly true in the unique components area, where the state of the art and quantity production may make drastic changes in the cost in the interim between the date of the estimate and the time the purchase order is written.

The above variables, plus many other intangibles, have resulted in the need for both a range of estimates composed of minimum, target, and maximum, and a method for evaluating the total presentation. An adequate method is outlined in this chapter.

AN EVALUATION TECHNIQUE

This method of evaluation starts with the assigning of a target cost, a minimum cost, and a possible maximum cost to each element of the program and to each subunit of the product. When these three sets of figures (three choices) are presented to the responsible management group, each manager starts to evaluate his immediate problems and try to project their effect on the anticipated program.

Figure 7-1 is representative of how a typical program can be presented. This summary is given a shakedown in a conference and a choice is made, based on various trade-offs and forecast conditions. The reason for each choice is documented and the chosen items of cost circled and totaled for a new target figure. This assures better estimating and promotes teamwork; there is no longer an estimating credibility gap, for the estimate has been changed to that of the managers.

A manager who has a voice in choosing a realistic target figure is likely to have a high morale level. Even top management can feel

ITEM	Min. Est.	Target	Max. Est.
1	(3.5k)	4.0k	5.0k
2	(3.0k)	3.0k	3.0k
3	9.0k	11.0k	(13.0k)
4	1.5k	(2.0k)	2.0k
5	5.5k	(6.0k)	6.5k
6	4.0k	4.1k	(4.2k)
The Total Equals - - - -	(31.7k)		

Fig. 7-1. Selective appraisal of item costs.

thwarted when they are handed only one voluminous estimate, and there is no possibility of making some desired change because to do so would scrap much of the work and there is not time to do the necessary rework.

This type of evaluation is recommended where the proper intelligence can be enlisted to use informed judgment. However, this should not be construed to minimize the importance of quoting a higher figure that would possibly be closer to the 80% probability zone.

PROBABILITIES AND BUSINESS RISKS

Probabilities so greatly affect business ventures that they should be of great concern in the presentation of any contract quotation. In Fig. 7-1 a total of $31,700 was chosen as a new target estimate. The sum of the minimum estimates is $26,500 and the sum of the maximum estimates is $33,700. The confidence level for the maximum estimate is definitely not 100%, but more likely in the area of 90%, and the low figure is not an impossible figure but possibly at the 35% confidence level. Then, the 90% and the 35% can be used to calculate the confidence level of the $31,700 target estimate, and to determine if it is advisable to use a higher figure, closer to the 80% confidence level.

A brief review of the geometry involved is always in order when one is dealing with as many varied problems as concerns the estimator. If all of the possible combinations and values for Fig. 7-1 from a confidence level of 0.01% to 99.99% were represented by straws and stacked according to their variation in length, we would have a stack of straws similar to Fig. 7-2, and the plane ABC, perpendicular to the straws would be considered a normal distribution curve as in Fig. 7-3. At the short end (A–A' of Fig. 7-2) there is but little possibility of doing the job for such a small amount, and at B–B' there is almost perfect assurance that the project would not exceed that amount.

Our minimum quote has been assigned 35% and the maximum quote 90%. These quotes are located on the plane $AA'BB'$ approximately as shown in Fig. 7-2, but we need a definite probability percentage scale for line A–B, and a scale B–B' on which we can post the estimated dollar amounts. As presented in Fig. 7-4, K & E's probability graph paper can be used to advantage in this case.

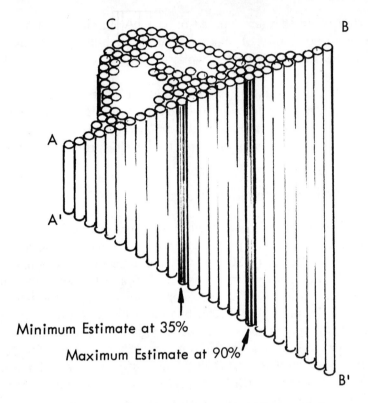

Fig. 7-2. All possible estimates between *A* and *B* represented as straws stacked according to length.

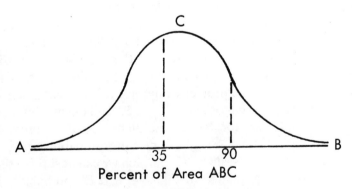

Fig. 7-3. Normal distribution curve.

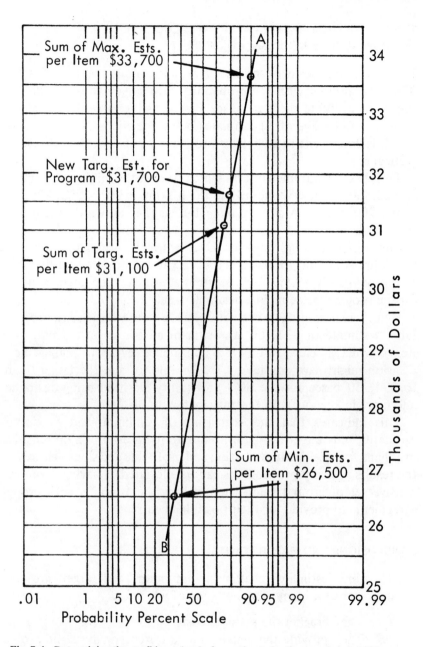

Fig. 7-4. Determining the confidence level of an estimate by the use of probability paper.

THE USE OF PROBABILITY PAPER TO CALCULATE
THE CONFIDENCE LEVEL OF AN ESTIMATE

The horizontal scale of probability paper similar to K & E's No. 358–23 is printed with the probability percentages given, so the first step is to calibrate the vertical scale by assigning values to it that will cover the full range of all estimates being considered, as in Fig. 7-4, which is equivalent to a sheet of probability paper with an elongated vertical scale.

This gives us a framework on which to post the minimum estimate of $26,500 at the 35% point and the maximum estimate of $33,700 at the 90% point. A straight line (A–B) is then drawn through these two points.

The revised target estimate of $31,700 for the program is noted on the dollar scale and followed across to the left to intersect line A–B at approximately 79.5%. This substantiates the revised target estimate for the program as being in a desirable area.

Had there been no item evaluation, there would have been a program target estimate of $31,100 (the sum of all the target estimates per item), which is close to the 75% confidence level, but management decision might have considered it advisable to raise the target figure to $31,750 based on the 80% probability position being desirable, and more in keeping with the anticipated problems.

This indicates that the estimating of minimum and maximum estimates, plus target estimates, tends to focus attention on sensitive problems for intelligent evaluation and profitable decisions. However, the results of many probabilities can be alleviated by having a comprehensive set of ground rules when making the estimate, with proper agreements to provide for their possible occurrence.

CONCLUSIONS AND MAJOR POINTS

1. The three estimates, maximum, minimum, and target, are very important for many reasons, some of which are:

   1.1. They bracket the possible costs.
   1.2. They provide the elements used for cost analysis and probability studies.

1.3. Cost analysis makes estimating a joint venture, bringing to light otherwise obscure problems.

1.4. Morale in the management circles is improved when each manager has a voice in the estimating of his unique task.

2. The best type of cost analysis is made by intelligent consideration of the possible costs and capabilities at the major task levels.

3. Probabilities are deceitful and one cannot build sufficient contingency allowance into a bid to guarantee 100% safety.

4. The documentation of the basis of an estimate, including comprehensive ground rules and supporting logic for controlling decisions, cannot be overemphasized.

5. A cost analysis and probability study that are well documented are also of prime importance.

8

The Basics of Learning Curves

The purpose of learning curves is to document the time (or cost) reduction experienced during the production of a series of like items; these experience curves can then be applied in the estimating and planning of similar programs. A thorough knowledge of learning curves and related techniques can be one of the most important assets of an estimator, production manager, or executive.

Even though the principles are simple and logical, learning curves are often applied in such a manner as to be a disadvantage to the user. This chapter starts with the rudiments in order to clearly define learning curves and bring more understanding to their application, and thus promote good practice.

The application of learning-curve techniques includes:

1. Strict adherence to comparing like programs as illustrated in Figs. 1-5 and 1-6.
2. Recognition of the various learning-curve theories, and the selection of the one best suited to the problem. (See Figs. 8-5 and 9-1.)
3. Construction of learning curves of any slope on log-log paper. (See Fig. 8-7.)
4. Construction of learning-curves through established cumulative average or unit amounts as plotted on log-log paper. (See Fig. 8-8.)
5. Determination of the slope of the learning-curve indicated by a number of datum points. (See Fig. 9-6.)
6. Forecasting of the revised cost due to cessation of production for a period of weeks or months. (See Fig. 9-5.)

7. The use of learning-curve tables for precision work. (See Fig. 8-8.)

8. Knowledge of learning-curve mathematics and the use of the pocket calculator to solve the formulas rather than use learning-curve tables. (See Chapter 10.)

THE EVOLUTION OF LEARNING CURVES

It is well to begin this review with the typical experience of a skilled worker on a small quantity of items as documented by the following schedule:

Unit	Unit Hours	Unit Hours X 2	Unit Hours X 3
1	10.0	20.0	30.0
2	8.0	16.0	24.0
3	7.3	14.6	21.9
4	6.3	12.6	18.9
5	6.0	12.0	18.0
6	5.6	11.2	16.8
7	5.6	11.2	16.8
8	5.0	10.0	15.0
9	5.1	10.2	15.3
10	4.5	9.0	13.5
11	4.8	9.6	14.4
12	4.6	9.2	13.8

These unit hours are plotted on arithmetical grid paper as in Fig. 8-1, curve *A*. *This is a learning curve.* Any new program of a similar nature can be expected to follow a relatively similar line.

In applying these data to another program, unit one at 10 hr could be considered unity and all other units would then be a certain percent of unit one; this is the basis of learning-curve tables.

In engineering work it is often necessary to attack a problem graphically and then to check mathematically for precision. This approach has many advantages in dealing with the application of learning curves, and should therefore be a firm rule in all learning-curve applications.

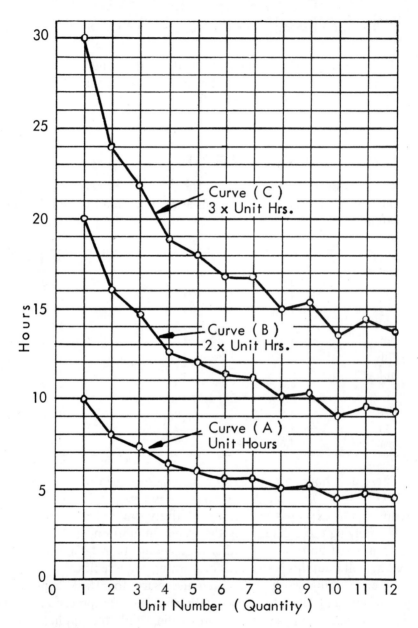

Fig. 8-1. Typical learning curve on arithmetical grid. (Note the change in shape when the curve is displaced vertically by factors.)

It should be noted how the unit curve changes shape on arithmetical grid paper as factors displace it vertically as in Fig. 8-1, curves *B* and *C*. This makes the use of templates impractical on arithmetic grid paper.

CHARACTERISTICS OF LEARNING CURVES ON SEMILOG PAPER

The logarithmic scale has been used to great advantage with learning curves. This is the same type of scale that is used on a slide rule. A typical arithmetical scale has equal divisions and the numbers start with zero. The logarithmic scale starts with unit one, and the units are progressively reduced, maintaining the unit size for any cycle at one-tenth the size of the previous cycle. This results in each cycle being similarly divided but having the numbers in each cycle ten times those in the previous cycle (note the hours scale in Fig. 8-2).

Figure 8-1 was used to plot the experienced hours on arithmetical grid paper; then the hours were factored by two and three to see what the resulting curves would be. These same data can be plotted on semilog paper for a more practical advantage as in Fig. 8-2, resulting in identical curves at all levels.

This makes it possible to cut a template for this curve and apply it graphically to other similar programs. Additional templates can be made as required for other types of work that follow different curves.

Many things can be accomplished with the use of semilog paper for plotting learning-curve problems. It is therefore well worth the effort to become familiar with this type of grid paper. Note that any specific percentage increase on the vertical axis (logarithmic scale) is always the same length in inches for any hourly amount. A 100% increase for the 10 hr of the first unit equals 20 hr, and the 100% increase of the 4.6 hr of the twelfth unit equals 9.2 hr, with the same distance between 10 and 20, in inches, as between 4.6 and 9.2.

Various programs can be readily compared on semilog paper.

CHARACTERISTICS OF LEARNING CURVES ON LOG-LOG PAPER

Changing the vertical axis to a logarithmic scale provides several advantages over the use of arithmetical grid, and making the horizontal scale (for quantity) logarithmic provides additional possibilities.

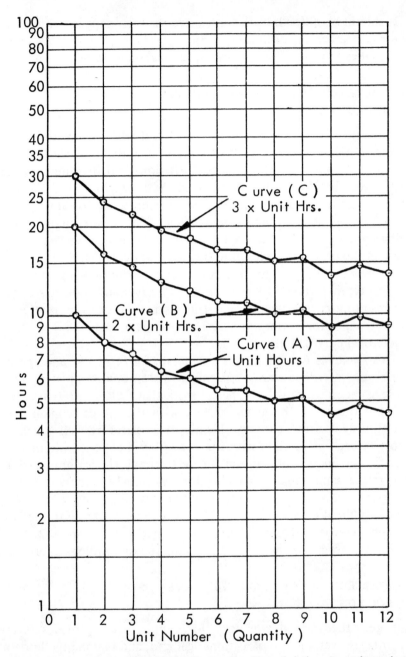

Fig. 8-2. Typical learning curve on Semilog paper. (Note that there is no change in the shape of the curve when it is displaced vertically by factors.)

Paper having logarithmic scales for both vertical and horizontal axes is called log-log paper. It is available in several sizes and configurations of 1 X 1 cycle, 2 X 3 cycle, and 3 X 5 cycle (divisions 1 through 10, and 10 through 100, or 100 through 1000, etc., are cycles). A 6-in. slide rule provides a ready scale when commercial paper is not at hand.

Plotting the same data as used in Figs. 8-1 and 8-2 on log-log paper results in a display such as shown in Fig. 8-3.

The obvious advantages in plotting a learning curve on log-log paper are:

1. The curves are identical at all levels and can be represented by a straight line, no template being required.
2. Vertical displacement of the curve when multiplied by a factor does not change the curve or the angle of an equivalent straight line drawn through the data.

With log-log paper, the vertical scale is used for hours or dollars, and it can be adjusted to the range of the estimate by multiplying or dividing the scale by some multiple of ten. However, the horizontal scale used for quantity should always start at unit one, and the theory of learning curves requires that this scale be for product units only.

One disadvantage in the use of log-log paper is the difficulty in distinguishing between unit values after the fortieth unit, but semilog paper can be used for an auxiliary chart in cases where this is critical.

THE THREE BASIC THEORIES

Thus far this chapter has dealt with the unit curve only, but there are two auxiliary curves, namely, the cumulative total and the cumulative average. Any point on the cumulative total curve is the sum of all the unit values through that point, and the cumulative average curve indicates the results of the cumulative total values divided by their respective quantities.

These two additional curves are plotted relative to the unit curve data in Fig. 8-4. The cumulative total line would have run off the chart on arithmetical grid paper; on semilog paper it would have been a convex curve, but on log-log paper it appears linear.

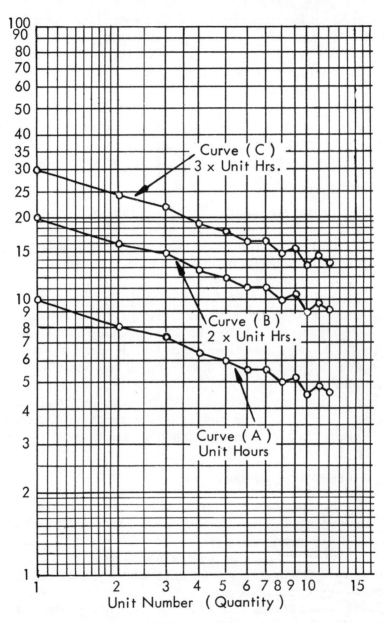

Fig. 8-3. Typical unit learning curve on log-log paper. (Note that there is no change due to the factors displacing the curve vertically.)

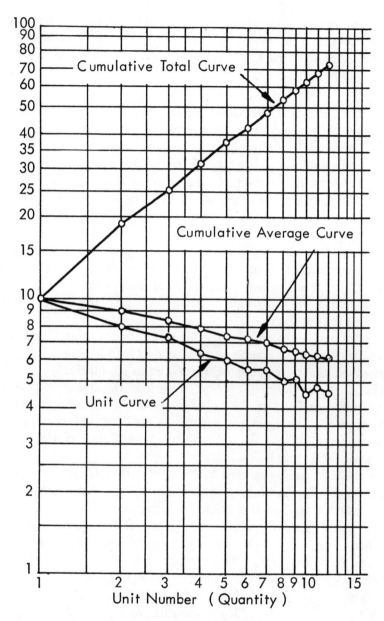

Fig. 8-4. Typical data used in the application of learning curves plotted on log-log paper.

The cumulative average and unit curves also seem to be linear, converging at unit one, but it will be found that they approach being parallel after the first few units. This presents three alternatives as illustrated in Fig. 8-5.

1. The unit line can be considered linear and the cumulative average will then be a convex curve for the first few units.
2. The cumulative average can be considered linear and the unit curve will be a concave curve for the first few units.
3. The cumulative average can be considered a slightly convex curve and the unit curve will be slightly concave, with unit one midway between the projection of the parallel lines approached by the two curves.

Any one of these theories is valid in cases where it is the curve of best fit. In brief, history attributes the honors for the first use of learning curves to Ted Wright. The earliest curves used were possibly the linear unit version.

The linear unit type of curve is often referred to as "the Boeing curve," Boeing having published linear unit learning-curve tables for in-plant use that have subsequently leaked to the outside.

The linear cumulative average version is the most practical for many purposes because the cumulative total is also linear, and the mathematics is simple in comparison to the linear unit version. This type of curve is often called "Northrop curves," since Northrop, too, published tables that leaked out.

The third version was used extensively at first, for it was war time when James R. Crawford wrote a manual for Lockheed Aircraft Corporation including this method, and gave night lessons in which its use was presented. Estimators and executives flocked from some distance to register and purchase a copy of "Crawford's Book" on this new learning-curve craze. This method has since fallen by the way, but those students received some very sound instruction.

It is well to keep in mind that there are at least two major learning-curve theories with their firm proponents, and in any discussion relative to learning curves one should first determine the specific theory under consideration, for their respective formulas differ. Starting

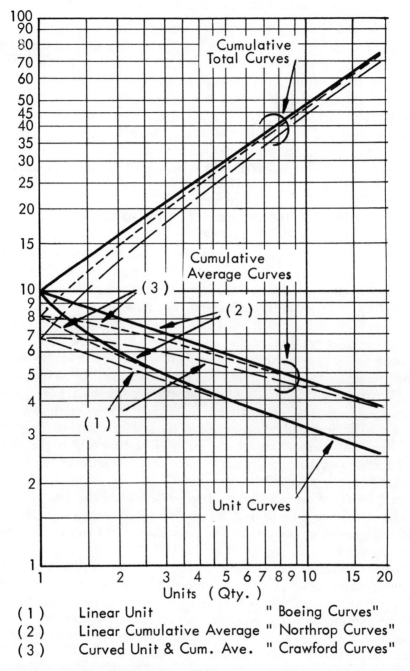

Fig. 8-5. The three major types of learning curves: (1) linear unit ("Boeing curves"), (2) Linear cumulative average ("Northrop curves"), (3) Curved unit and cumulative average ("Crawford curves").

with the same unit one value, the projection of additional units will be different, and if several datum points after unit 20 are used as a base, each will result in a different unit one value (see Fig. 8-5).

The major differences in the two theories occurs in the first few units, so in cases where accounting is by lots there may be no way of telling which theory provides the curve of best fit.

When one has mastered one method the others are easily understood, so the remainder of this chapter will deal with the linear cumulative average theory.

CONSTRUCTION OF LINEAR CUMULATIVE AVERAGE LEARNING CURVES ON LOG-LOG PAPER

Learning-curve techniques are applicable to a whole family of curves that follow the various experienced improvement rates, and they are defined by the percent of improvement that occurs each time the cumulative total quantity is doubled. For instance, an 80% learning-curve is indicated when the first ten items require 100 hr each, and the average for the first ten plus an additional ten units is 80 hr. (The second group of ten must be done in 60 hr each to average 80 hr for the cumulative total of 20 units, not 80 hr each as is often assumed.)

Constructing the linear cumulative average line on log-log paper for any desired percentage curve is the basis of making learning-curve charts. The best method makes use of the above definition: An 80% curve is easily drawn through any point by plotting 80% of the given value at double the quantity at the given point. This can be continued for additional points and greater accuracy.

A protractor can also be used for determining the slope of the cumulative average line, for each curve has a specific angle as indicated by the tangent to the slope. The respective curves and their angles are given in Fig. 8-6.

Besides being a table of angles that can be used in constructing typical percentage curves on log-log paper, Fig. 8-6 provides the relationship of the unit curve to the cumulative average at sufficient points for one to use a French curve in drawing the curved unit line. The cumulative total line is drawn after the cumulative average line has been constructed. Unit one and 100 times the cumulative average at unit 100 can determine the two points needed for constructing the cumulative total curve, which is linear on log-log paper.

Curve %	Angle on Log-log Paper		– R – VALUE Ratio, Unit Amount to the Cumulative Average			
			At Unit 2	At Unit 4	At Unit 10	At Unit 100 & up
99	0°	50′	.990	.987	.986	.986
98	1	40	.980	.975	.972	.971
97	2	31	.969	.962	.958	.956
96	3	22	.958	.949	.944	.941
95	4	14	.947	.935	.930	.926
94	5	6	.936	.922	.915	.911
93	5	59	.925	.908	.900	.895
92	6	52	.913	.894	.885	.880
91	7	45	.901	.880	.870	.864
90	8	39	.889	.866	.855	.848
89	9	32	.876	.851	.839	.832
88	10	27	.864	.837	.823	.816
87	11	22	.851	.821	.807	.799
86	12	17	.837	.806	.791	.782
85	13	12	.824	.791	.775	.766
84	14	7	.810	.775	.758	.748
83	15	3	.795	.759	.741	.731
82	15	59	.780	.742	.724	.714
81	16	55	.765	.726	.707	.696
80	17	23	.750	.709	.689	.678
79	18	47	.734	.692	.672	.660
78	19	43	.718	.674	.654	.642
77	20	40	.701	.656	.635	.623
76	21	36	.684	.638	.617	.604
75	22	33	.667	.620	.598	.585
74	23	29	.649	.601	.579	.566
73	24	25	.630	.582	.558	.546
72	25	21	.611	.563	.539	.526
71	26	18	.592	.544	.518	.506
70	27	14	.571	.524	.499	.485

Fig. 8-6. Table for constructing learning curves on log-log paper.

CONSTRUCTION OF A TYPICAL 80% LEARNING
CURVE ON LOG-LOG PAPER

Given: 50 hr as the average time for the first five units of a program.

Problem: Construct an 80% learning-curve chart for the continuing program.

Method: Use the data for an 80% learning curve as given in Fig. 8-6.

A protractor can be used to draw the cumulative average line at 17° 23' through 50 hr on the unit five line as given in Fig. 8-7. (The alternate method is to multiply 50 hr by 80% to equal 40 hr cumulative average through unit ten, and 80% of 40 hr equals 32 hr through unit 20 to provide three points for accuracy.)

The unit line is at 0.750 of the cumulative average for unit two; 0.709 of the cumulative average at unit four; 0.689 of the cumulative average at unit ten; and 0.678 of the cumulative average at unit 100.

Percentages are a constant distance on logarithmic scales as noted in Fig. 8-7, so these factors are distances easily picked off the chart by dividers or by tick marks on a bit of note paper and then transposed to the respective units to determine the unit curve.

The text format in Fig. 8-7 stops short of unit 100, but the cumulative total line can be plotted at 20 times the cumulative average through unit 20. It should also be noted that the distance a is the same as the distance a'; this is true for all slopes.

For most practical applications the readings taken from a well-constructed chart are adequate, but some conditions dictate the need for more apparent precision, or consistency, which is obtained through the complementary use of learning-curve tables, or direct calculations that can be made on a pocket calculator as given in Chapter 10.

LEARNING-CURVE TABLES AND THEIR APPLICATION

Learning-curve tables are compiled so that the first item is always unity (or 100%) and all other units are given in relative decimal fractions (or percentages). Figure 8-8 duplicates the data and curve given in Fig. 8-7. This is done to emphasize the need for a chart even when learning-curve tables are used for determining any needed value.

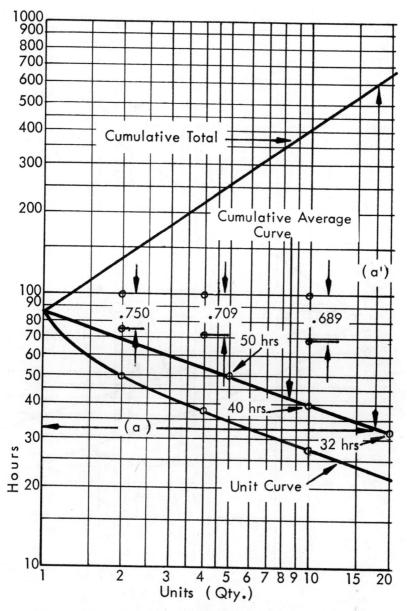

Fig. 8-7. Constructing an 80% linear cumulative average learning-curve chart.

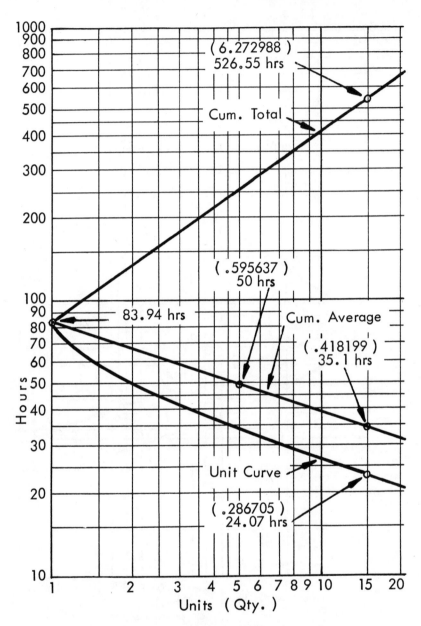

Fig. 8-8. Learning-curve tables used in conjunction with charts.

The first operation after completing the chart is to turn to the proper tables (80% in this case) and copy the table figures for the given value. For the cumulative average at unit five it is 0.595637. The cumulative average for the first five units (50 hr) is then divided by the table value (0.595637) to obtain the hours for the first unit (83.94 hr).

Learning-curve forecasts can now be made for the quantity of 15 items as given in Fig. 8-8. The learning-curve table values are noted on the chart and then used as factors of the unit one amount to obtain the cumulative average through 15 as 35.1 hr, the cumulative total of 526.55 hr, and the fifteenth unit at 24.07 hr. The next important operation is to verify the mathematical results with the readings from the chart.

Learning-curve tables are impressive, being extended to six or more decimal places, but much of this apparent accuracy is wasted because it is seldom that one does not have to choose, for example, between an 81% or an 82% curve when the trend is somewhere between the two slopes. With the advent of the pocket calculator it is now possible to make the needed calculations for the precise curve indicated by the data.

CONCLUSIONS AND MAJOR POINTS

1. There are several types of learning curves, so the cognizance of which is being presented is of the highest priority.
2. Learning curves are twice as critical as often thought. An 80% curve means that the second lot of equal size must be done for 40% less than the first.
3. The purpose of log-log paper is to provide for linear projection, and any use of curves on log-log paper is usually self-defeating.
4. One should compare like programs; prototype data to prototype production; preproduction data to preproduction; and production data to production.
5. There are a number of methods used in determining the slope of a learning curve, but they are all based on the two methods given in this chapter.
6. Learning-curve tables are very helpful in conjunction with the charting of learning curves for they provide the equivalent of a more narrow line.

7. The larger the projection forecast quantity, the greater the risk.

8. Minimum learning-curve tables are included in Appendix 2.*

---

*The most comprehensive learning-curve tables are published by:

Western Periodicals Co.   Phone (213) 875–0555
13000 Raymer Street
North Hollywood, California 91605

These tables are for both the linear cumulative average and the linear unit types, from 51% through 99% for all units, one through 2000.

9
Extended Learning Curve Application Techniques

Chapter 8 provides the basics and good practice techniques in the use of learning curves, but their actual application often extends far beyond these basics. Learning curves must not be permitted to become a trap, for they are an important tool to be used in assisting the estimator or manager in using his judgment consistently. The following problems occur frequently, but they have logical solutions that can be easily applied when the need arises.

The type of learning curves experienced by a particular firm will change with the plant experience and the degree of automation introduced, for there is a carry-over from one program to the next. A program affected by almost any drastic change, such as new or improved production conditions, can best be charted as if it were the beginning of a new program; one should separate the previous data and start back at unit one on the chart with the first serial number affected.

Cumulative average curves are usually the only practical way of documenting learning (progress); such curves are very often misleading, for the unit curve can actually be increasing when the cumulative average is still indicating a downward trend. It is therefore necessary to plot a synthetic unit curve in order to interpret the trend correctly.

The regression in learning that occurs during a period of nonproduction has long been a problem, but there is a rather simple method of handling this condition. This formula is also applicable to follow-on programs where the continuity is broken.

The line of best fit cannot be correctly drawn on log-log paper by the "least squares formula" through a group of data having a broad scatter; the averaging method is much better in such cases. (See Fig. 9-6.)

THE PROGRESSION AND SIMPLIFICATION OF
VARIOUS TYPES OF LEARNING CURVES

A progression in the types of learning curves experienced is not uncommon. These phases can be identified and their subtle differences utilized. Constant improvement in tools and techniques produce the initial curve, which is linear on log-log paper, as in Fig. 9-1A. This curve is shown as plotted on both arithmetic grid paper and log-log grid paper.

After the tooling and techniques are established the curve may become more complex. The initial items will start as in the beginning, then there will be an impressive drop as the advanced tooling is introduced as shown in Fig. 9-1B.

Automated and semiautomated equipment may be the next step, which produces a "broken curve" such as Fig. 9-1C. The initial items will start as at first and drop drastically with the use of some automated equipment; then the change to the maximum automated phase results in a flat curve.

The humped curve as in Fig. 9-1D can very well follow any one of these first three phases, as the next step is to use production equipment from the start, if it is flexible enough. This type of curve is known as the "Stanford curve."

Each one of these curves, except Fig. 9-1A, has departed from the basic purpose of log-log paper, that of linear projection. Estimating the curve variations to reflect forecast conditions can become quite complex with any one of these experienced curves when handled in the conventional manner. However, there is one simple technique that will satisfy all cases and simplify the data presentation. Curve A is linear and therefore there is no problem, but curves B, C, and D are shown in their improved and simplified manner in Figs. 9-2, 9-3, and 9-4.

The Dog-Leg Curve Simplified

This type of curve is very common with firms that have been producing a certain type of complex item for three to ten years. It results from the initial quantity being produced on soft tools and a follow-on quantity being produced on hard tools with techniques and learning from the previous programs being suddenly introduced.

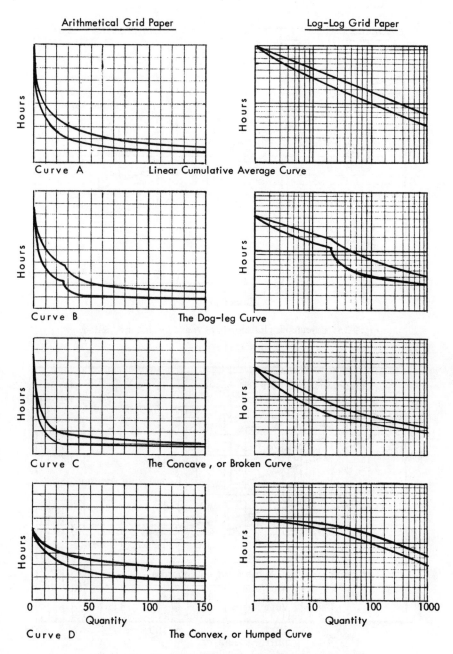

Fig. 9-1. Various types of learning curves illustrated on both
arithmetical grid and logarithmic grid.

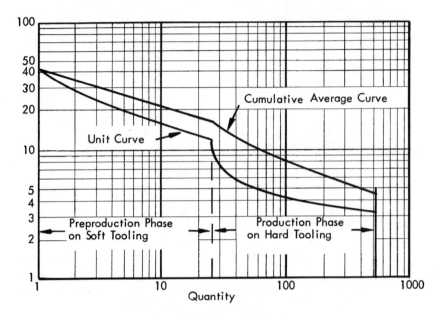

Fig. 9-2A. Cumulative plotting of preproduction and production.

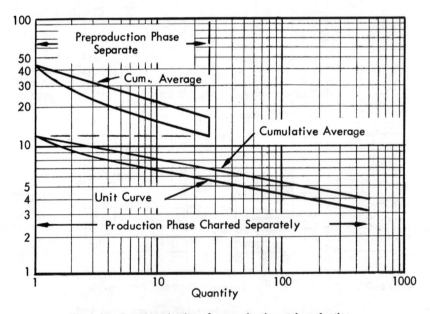

Fig. 9-2B. Separate plotting of preproduction and production.

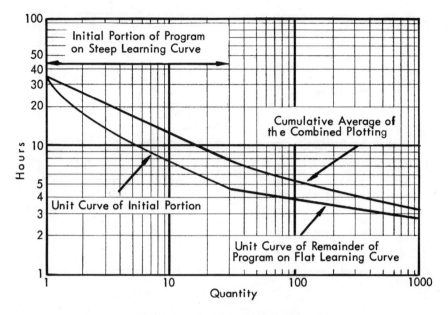

Fig. 9-3A. Cumulative plotting of data forming broken curve.

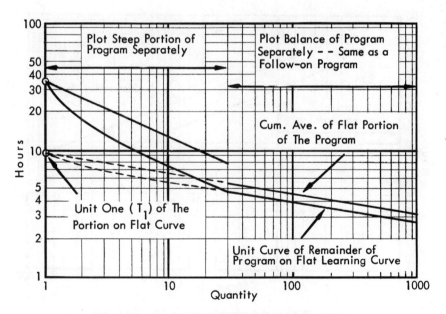

Fig. 9-3B. Separate plotting of data used in Fig. 9-3A.

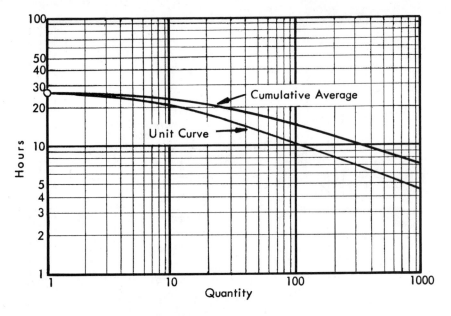

Fig. 9-4A. The convex, or humped, curve.

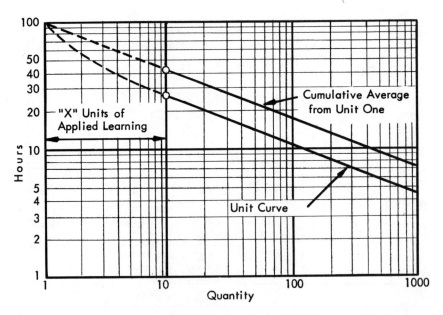

Fig. 9-4B. The humped curve plotted as a follow-on program.

There is no problem with this type of curve when one understands that to plot these two phases in one continuous curve results in unrelated programs being run together as in Fig. 9-2A. Figure 9-2B shows how simple the problem can be made if the two programs are separated. Future work can be easily forecast with the use of normal linear techniques as per the initial concept for using log-log paper.

There are a number of major changes in conditions that can distort a continuous plotting of the progress curve, making the curve apparently useless for forecasting a follow-on program. Such cases can be solved, or greatly simplified, by dividing the program after the last unit prior to the change and plotting the continuation of the program under the new conditions as a separate entity.

A move to new facilities or extensive new tooling in the midst of a program may need to be handled in this manner. A restart after a long period of nonproduction is often best handled thus. Extensive reengineering is another reason to start over at unit one.

There is no other one method that adds as much latitude to the application of learning curves as this, for it covers all problems that have created the apparent need for various curves on log-log paper.

The Broken Curve Simplified

This type of curve is experienced by firms producing very complex items, but they have developed tools, processes, and automated equipment to assist in much of the exacting work. This results in a very fast rate of learning on the first 30 or so items; then the curve flattens out to a curve in the 90% plus area.

The data for these curves are often cumulative average figures, with no lot figures available to indicate the unit curve trend. There is little to oppose the assumption that follow-on programs will continue on a curve tangent with the last few cumulative average figures. In such cases one can graphically analyze the cumulative average data by using a French curve to make a representative smooth cumulative average line. It is then possible to calculate sufficient unit values so that a representative unit curve can be drawn. The unit data points can be calculated by reading the cumulative average figures for approximately every fifth unit, calculating the midunit value for each group of five, and plotting each at the proper midunit. The unit line is then drawn through these points, using a French curve to smooth out the line. This type of graph presents the level and final slope of the unit curve, providing clarification of the problem.

This problem is solved in the same manner as used in Fig. 9-2 for the dog-leg curve. Everything falls into place if the initial quantity on which there is steep learning is plotted separately, and then the settled production of the following items will be linear on log-log paper. This provides reliable linear forecasting for follow-on contracts and other needs. Figure 9-3 illustrates such a problem.

The Humped Curve, or Stanford Curve, Simplified

Every variation to the learning-curve theory has been developed to meet some particular problems when they have risen. Repeating programs will each have a great affect on the following programs. Projects that are not highly complex in which the same or similar tooling is used can well start at a low point, which is equivalent to starting a program after a small initial quantity has been produced.

In using this theory to greatest advantage one must be able to use learning curves of any percentage and to start easily at any unit, be it two or more. This can be done by handling this type of program in the same manner as a follow-on program. Any amount of previous learning can then be applied to any new program, following a curve of any desired percent.

It has been stated in jest that the first five units are too expensive, so the plan is to start with the sixth unit. This is a fine idea if it is feasible. Such an idea is often a temptation that must be ignored unless there is previous history on very similar programs to support the plan.

Figure 9-4 illustrates this type of program with two charts. The first (A) shows the humped curve on log-log paper as it develops when started at unit one, and (B) shows how the task is simplified by treating the project as one would a follow-on program.

RESTART SETBACK

Retained learning is greatly affected by a termination and restart of a program, elapsed time between lots, or elapsed time between programs. This is reflected in the learning curve. Many authorities concede that there is a regression of learning during the period of nonproduction, and that the restart effort should be assumed to start at some unit less than the last serial number produced at the termination. However, due to the lack of a formula, the number of units for this setback is left to judgment alone.

The fact that it requires the same amount of foot-pounds to blast a missile straight up for 1000 ft as the same missile would generate if it dropped from the same height suggested the following empirical formula, which can be checked by experience and adjusted to conform to any type of work.

Formula: The amount of setback on the learning curve to locate the first unit of a restart is equal to the rate of production per month at the time of termination multiplied by the number of months that the program was suspended.

Illustration: Assuming a production rate of ten units per month at the time of cessation of production, and the program being halted for two months; then the setback should be 2 × 10 to equal 20 units as shown in Fig. 9-5. This type of formula can assist one in using his judgment and provide for any adjustment that may seem proper.

An impressive paper containing estimating formulas and their use came to my desk one day, and the table of contents listed a formula for determining the restart setback. This I could use! On turning to the proper page I found this notation: "We are still working on this one." This was the second time that I had found such a notation in estimating literature, and the shock and amusement were just what I needed at the time.

AVERAGING METHOD FOR LINE OF BEST FIT

The "averaging method" and the "least squares formula" are both mathematical methods of drawing the "line of best fit" through a group of datum points on arithmetical grid paper. Both methods can also be used with data plotted on log-log paper, but accuracy will suffer in the use of either method, for the real line of best fit is somewhere between the two results. This is most noticeable when the data has a wide scatter, varying greatly in the Y range.

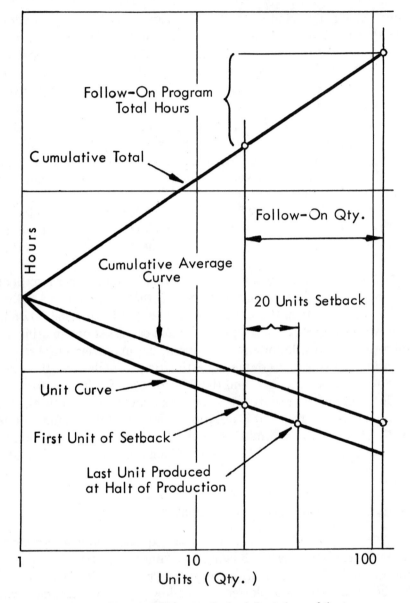

Fig. 9-5. Formula for restart setback relative to lapse of time.

Locating the learning curve by the averaging method may result in a position that is slightly higher than the true line of best fit for the initial quantity and be lower than the line projected by the least squares method for a follow-on quantity. Conversely, the least squares formula when used on log-log paper may misplace the line to a low inaccurate position for the initial quantity when the scatter is broad, and project a figure for the follow-on quantity that is higher than that projected by the averaging method.

For the averaging method, the data are first divided into equal groups; then the average of all the dollars (or hours) is determined separately for each of the two groups. The midpoint of each group is then determined by averaging the unit positions of the data within each group. The midposition of each group, with its average when posted on a chart as in Fig. 9-7, provides two points to determine the line of best fit by the averaging method.

An example: Fig. 9-7 uses the same data as was used for Fig. 2-6. The costs are in even thousands, so we can ignore the last three zeros of the cost values.

	Group One		Group Two	
	Qty.	Amt.	Qty.	Amt.
	100	304	400	240
	150	320	530	240
	200	340	800	280
	450	964	1730	760
Averages	150	321.3	576.7	253.3

CONCLUSIONS AND MAJOR POINTS

1. Learning curves have many applications, and they will be found in many configurations.
2. The various phases experienced with learning curves are the results of changing conditions.
3. The several sponsored types of nonlinear learning curves on log-log paper were developed to account for unique phases in the progression of the curve, but they are for the most part self-defeating because of their complexity.

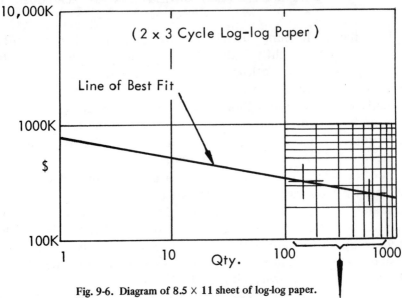

Fig. 9-6. Diagram of 8.5 × 11 sheet of log-log paper.

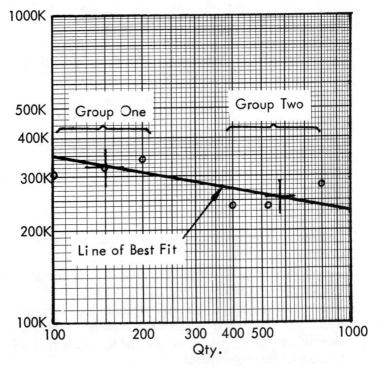

Fig. 9-7. Line of best fit by averaging method.

4. Any major change from the linear plot on log-log paper represents a major change, and it is best to make a break in the records and restart at unit one with the first unit affected by the major change.
5. A cumulative average curve alone provides very poor visibility for forecasting future costs, but it can be interpreted by the construction of a synthetic unit curve.
6. The restart setback formula to account for the regression in learning can be of great assistance in solving this puzzling problem.
7. The line of best fit through a group of data is not a problem to consider lightly, for it is the basis of the forecast, and it is wise to use both the least squares formula and the averaging method to determine which provides an advantage.
8. The least squares formula is rather involved for use with even a pocket calculator, so this operation is done by computers for the most part.

10

Pocket Calculator Assist for Learning Curve Calculations

In the past it was necessary to depend almost exclusively on learning-curve tables. This often led to problems, such as deciding whether a curve should be 86% or 87%. The buyer would insist on the 86% and the vendor on 87% when it should probably have been 86.5%. The electronic slide rule calculator can readily compute the 86.5% learning-curve factors. This can be important, for the difference between an 86% and an 87% curve is 8% at unit 100, 12% at 1000, and 13.5% at 2000.

The capability of the electronic hand-held calculator to provide directly the power of a number to a decimal fraction as required in learning-curve formulas is possibly the element of greatest importance in solving learning-curve formulas. This simplifies the use of formulas, reducing the time required, so whether to use tables or formulas can be a matter of choice.

However, it is not recommended that everyone who uses learning curves scrap the learning-curve tables and rely exclusively on the calculator, though this may be the easiest and best course for some.

The following are suggestions for choosing the type of calculator and simplifying the use of the calculator and formulas.

A ten-digit calculator with log and trig functions, plus a memory, are minimum requirements. There are many variations in the different makes; therefore, choose one that is best for you. One that is programable is an advantage, especially if one plans on using linear unit curve formulas (Boeing curves).

A small notebook that will slip into the calculator case is a necessity. Pertinent data on the curves most used should be entered, and there should be a page for each formula with a diagram of the routine for entering the formula into the calculator.

BASIC LEARNING-CURVE FORMULAS

The letters designating the various elements, plus the formulas used to calculate learning-curve values are as follows:

T = Time $\qquad$ X = Quantity or Units
T_1 = Time at Unit One $\qquad$ T_U = Unit Time
T_C = Cumulative Total Time $\qquad$ T_A = Cumulative Average Time
P = Learning-Curve Percent $\qquad$ K = Tangent to Slope
L = $(1 - K)$ Which is one minus the tangent.

A table of the values for the constants K and L are included (see Fig. 10-4), but the following formula is necessary to develop the values of K and L for learning-curve percentages not given:

$$K = (\log 100 - \log P) \div \log 2$$

Three basic formulas are used to calculate the various values given in learning-curve tables, or to directly calculate the respective values for a program. These formulas are shown in their proper relation to a diagram of the three curves used, and as they appear on log-log paper. (See Fig. 10-1.)

The formula values of X are easily raised by a power that is a decimal fraction, as this operation is automatically done by pocket calculators. Refer to the directions included with the calculator.

Routines and examples for each formula are presented later in this chapter. A Texas Instruments SR-50 Slide Rule Calculator was chosen as a minimum instrument for this type of work.

A little study of the instruction manual for the respective calculator chosen should enable one to diagram compatible routines for each formula.

THE KEY TO THE MATHEMATICS OF LEARNING CURVES

There is a key to comprehending the mathematics involved in learning-curve calculations: If one had a stick 10 in. long with random marks on it, the location of each one of the marks could be determined by

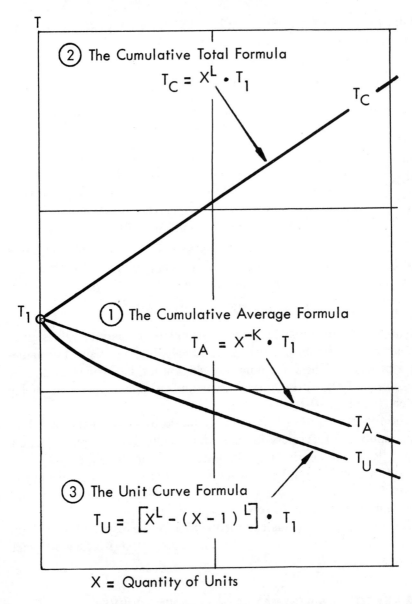

Fig. 10-1. The three basic learning-curve formulas: (1) the cumulative average formula;
(2) the cumulative total formula; (3) the unit formula.

placing a 10-in. ruler beside the stick and noting the number of inches to each mark. This is approximately the procedure necessary for much of the mathematics used with learning-curves. Each number from the chart on log-log paper must be converted to a measurable value. The log scale cycle length provides this measurable unit value, and the log of any number that appears on the log-log paper converts it to this measurable scale. This is the relationship of the arithmetical values (or scale) used for the simple mathematics and the logarithmic scale of the paper. Thus, cycle 1 is unit one; cycle 2 is unit two, etc.; and the mathematics considers the decimal fraction value of each number between these whole numbers.

The location of any number on the logarithmic scale between one and ten can be converted to zero plus a decimal fraction; between 10 and 100 each is converted to one plus a decimal fraction; and between 100 and 1000 each is converted to the digit two plus a decimal fraction. This is illustrated in Fig. 10-2.

Finding the log of a number is a direct operation on the pocket calculator as noted in instruction books. One needs but determine the log of each datum value and proceed with the problem in a simple mathematical manner. The result will then sometimes require finding the antilog; for this there is a simple procedure given in pocket calculator instruction manuals.

The Tangent K

One of the basic elements of learning-curve mathematics is the tangent for the angle of the curve slope. It is the K in learning-curve formulas which is a constant relating to each different curve percentage.

A simple definition of the tangent is: The ratio of the short side of a right triangle to the long side. See Fig. 10-3(A), in which the triangle has been inverted to indicate the negative slope of a learning curve. The tangent of angle a is $A \div B$.

In Fig. 10-3(B) the values for an 80% curve are noted at each corner of the triangle. It is necessary to obtain the log of each one of these figures to know the length of the two sides of the triangle. These are tabulated thus:

Log of 1 is 0   Log of 80 is 1.9031
Log of 2 is 0.30103   Log of 100 is 2.0

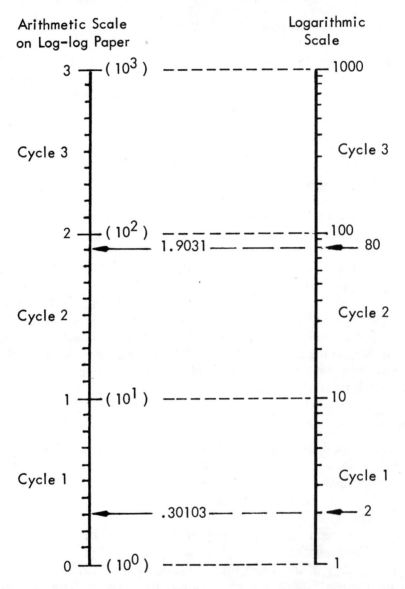

Fig. 10-2. The two scales on log-log paper: The arithmetic scale is used for mathematical work. Note that the power of 10 in parenthesis is the same as the respective cycle number.

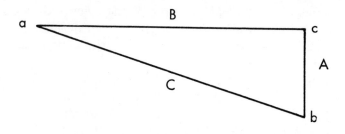

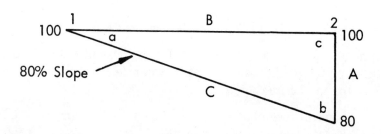

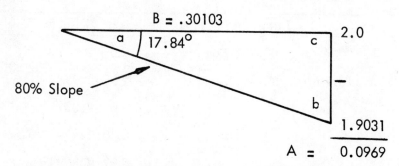

Fig. 10-3. Developing the tangent K for an 80% learning curve—typical procedure.
(A) Arithmetic tangent—$A \div B$. (B) Data for 80% learning curve on log-log paper.
(C) Log values of 80% learning curve used to determine the tangent.

This is illustrated in Fig. 10-2. There is one element of a logarithmic scale that can be confusing: The logarithmic scale actually contains the values of the antilog numbers, and the arithmetic scale contains the log numbers.

These log numbers are added to the triangle in Fig. 10-3(C). The log of 2 is the dimension of side B, for the log of one is zero; but the length of side A is obtained by subtracting the location of b at 1.9031 from the location of c at 2.0 to equal 0.0969 the length of side A. Then the tangent of a is $A \div B$, or $0.0969 \div 0.30103$ to equal 0.32189, the tangent for 17.84°. See your calculator instruction manual for the method of determining the angle from the tangent.

Calculating the Tangent K, the Angle, and the Constant Ratio L

The next step is to note the formula for the tangent K and develop a routine for the pocket calculator that will result in the tangent and the angle for any desired slope.

Tangent formula: $K = (\log 100 - \log P) \div \log 2$
P is the learning-curve percent.

We can establish a routine by using the data of Fig. 10-3 and perform the necessary operations required for an 80% learning curve as follows:

	Enter	Press		Display
	100	Log	–	2.0
Curve Percent	80	Log		1.9030!)987
		=	÷	0.096910013
	2	Log		0.3010299957
		=		0.3219280949 for K
		Arc	Tan	17.845° for the angle

TABLE OF " K " AND " L " VALUES BY ONE HALF PERCENT

Curve Percent	Tangent " K "	(1 – K) " L "	Curve Percent	Tangent " K "	(1 – K) " L "
99.5	.00723157	.99276843	80.5	.31293931	.68706069
99	.01449957	.98550043	80	.32192809	.67807191
98.5	.02180437	.97819563	79.5	.33097323	.66902677
98	.02914635	.97085365	79	.34007544	.65992456
97.5	.03652588	.96347412	78.5	.34923544	.65076456
97	.04394335	.95605665	78	.35845397	.64154603
96.5	.05139915	.94860085	77.5	.36773178	.63226822
96	.05889369	.94110631	77	.37706965	.62293035
			76.5	.38646835	.61353165
95.5	.06642736	.93357264	76	.39592868	.60407132
95	.07400058	.92599942			
94.5	.08161377	.91838623	75.5	.40545145	.59454855
94	.08926734	.91073266	75	.41503750	.58496250
93.5	.09696173	.90303827	74.5	.42468767	.57531233
93	.10469738	.89530262	74	.43440282	.56559718
92.5	.11247473	.88752527	73.5	.44418384	.55581616
92	.12029423	.87970577	73	.45403163	.54596837
91.5	.12815635	.87184365	72.5	.46394710	.53605290
91	.13606155	.86393845	72	.47393119	.52606881
			71.5	.48398485	.51601515
90.5	.14401030	.85598970	71	.49410907	.50589093
90	.15200309	.84799691			
89.5	.16004041	.83995959	70.5	.50430484	.49569516
89	.16812276	.83187724	70	.51457317	.48542683
88.5	.17625064	.82374936	69.5	.52491512	.47508488
88	.18442457	.81557543	69	.53533173	.46466827
87.5	.19264508	.80735492	68.5	.54582411	.45417589
87	.20091269	.79908731	68	.55639349	.44360651
86.5	.20922796	.79077204	67.5	.56704059	.43295941
86	,21759144	.78240855	67	.57776700	.42223300
			66.5	.58857375	.41142625
85.5	.22600367	.77399633	66	.59946207	.40053793
85	.23446525	.76553475			
84.5	.24297675	.75702325	65.5	.60143319	.38956681
84	.25153877	.74846123	65	.62148838	.37851162
83.5	.26015190	.73984810	64.5	.63262893	.36737107
83	.26881676	.73118324	64	.64385619	.35614381
82.5	.27753396	.72246604	63.5	.65517150	.34482850
82	.28630419	.71369581	63	.66657627	.33342373
81.5	.29512804	.70487196	62.5	.67807191	.32192809
81	.30400619	.69599381	62	.68965988	.31034012

Fig. 10-4. Values of K and L to be used with learning-curve formulas.

This calculator routine covers the same operations performed for Fig. 10-3.

The second value used in learning-curve formulas is the constant ratio L, which is simply $(1 - K)$. For an 80% curve it is $1.0 - 0.3219280949$, or 0.6780719051.

This method is applicable for any slope, and it will be necessary to use this formula when the values of K and L do not appear in the tables of Fig. 10-4.

CUMULATIVE AVERAGE FORMULA AND POCKET CALCULATOR ROUTINE

The cumulative average learning curve is linear on log-log paper. These formulas and pocket calculator routines for use in learning-curve projection provide the needed learning-curve table value plus the desired amount for the problem. This is all done by one routine.

$$\text{Cumulative Average Formula: } T_A = X^{-K} \cdot T_1$$

Example—problem: Find the cumulative average amount (T_A) at unit X, which is 100, for an 80% learning curve when unit one (T_1) is 90 hr.

Figure 10-4 provides the value of the tangent K for an 80% curve as 0.32192809. Substituting these values in the formula:

$$T_A = 100^{-0.32192809} \cdot 90$$

Calculator routine:

Symbol	Enter	Press	Display
X	100	y^x	
K	0.32192809	$+/-$ $\times$	0.2270616666 (This figure is the learning-curve table value.)
T_1	90	$=$	20.43555 hr for T_A, the cumulative average

Figure 10-5 shows how these figures fit into the learning-curve graph. Note: This same formula is the unit curve formula for the linear unit type of learning curves.

CUMULATIVE TOTAL FORMULA AND POCKET CALCULATOR ROUTINE

The cumulative total learning-curve line is also linear on log-log paper.

$$\text{Cumulative Total Formula: } T_C = X^L \cdot T_1$$

Example–problem: Continuing with the same problem as for the cumulative average in Fig. 10-5: What is the cumulative total (T_C) factor at unit 100 for an 80% learning curve when the first unit (T_1) is 90 hr?

Figure 10-4 provides the value of L, which is 0.67807191 for an 80% learning curve. This is a good figure to remember, for it is often needed. Substituting the values in the formula:

$$T_C = 100^{0.67807191} \cdot 90$$

Calculator routine:

Symbol	Enter	Press	Display
X	100	y^x	100
L	0.67807191	X	22.70616666 (This is the learning-curve table value.)
T_1	90	=	2043.554994 hr for T_C, the cumulative total

Figure 10-6 depicts the problem through this step. Note that both the learning-curve value and the hours are posted on the graph. This permits the checking of the calculations by one who uses learning-curve tables only.

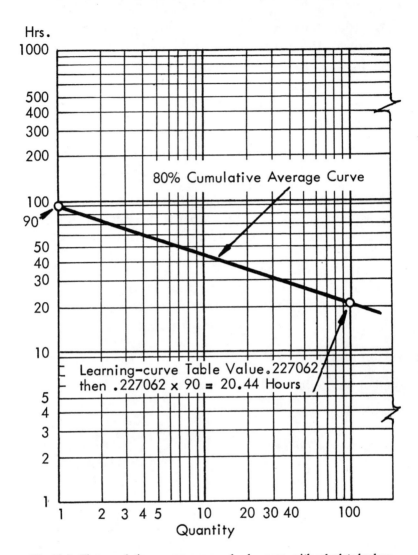

Fig. 10-5. The cumulative average curve on log-log paper with calculated values.

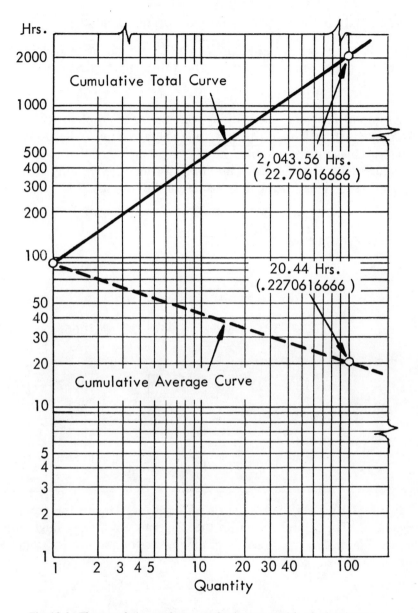

Fig. 10-6. The cumulative total curve on log-log paper with calculated values.

UNIT FORMULA AND POCKET CALCULATOR ROUTINE

The unit curve to be compatible with a linear cumulative average line on log-log paper will be convexly curved through the first few units.

$$\text{Unit Formula: } T_U = [X^L - (X-1)^L] \cdot T_1$$

Example–problem: Continuing with the same problem as for Figs. 10-5 and 10-6, calculate the unit time value (T_U) at unit 100 for an 80% learning curve when unit one (T_1) is 90 hr.

Figure 10-4 provides the value of L for an 80% learning curve as 0.67807191. Substituting these values in the formula:

$$T_U = (100^{0.67807191} - 99^{0.67807191}) \cdot 90$$

Calculator routine:

Symbol	Enter	Press		Display
X	100	y^x		100
L	0.67807191	Sto	$-$	22.70616666
$(X-1)$	99	y^x		99
L	(0.67807191)	Rcl	$=$	0.1542130629 $(T_U$ table value)
		X		
T_1	90	$=$		13.87917566 hr for T_U

Figure 10-7 gives the results of adding these data to the graph in Fig. 10-6. The figures shown in the diagrams are the readings from the calculator display; they are given for clarity and not to promote delusions of accuracy.

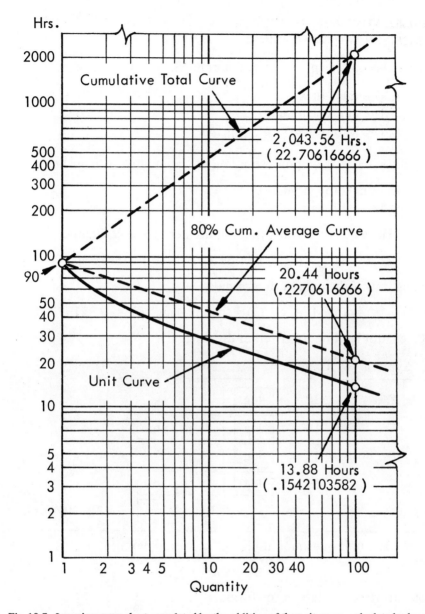

Fig. 10-7. Learning-curve chart completed by the addition of the unit curve—calculated values.

CUMULATIVE AVERAGE CURVE
BASED ON TWO GIVEN POINTS

Three basic elements must be determined:

1. The value of K.
2. The unit one amount (T_1).
3. The learning-curve percent (P).

Given data: Two points only on the cumulative average line.

a. 47 hr through unit 10.
b. 33 hr through unit 30.

1. The value of K is the first requirement to be calculated.

$$K = (\log T_a - \log T_b) \div (\log X_b - \log X_a) \quad \text{(See Fig. 10-8)}$$

Substituting the data in the formula:

$$K = (\log 47 - \log 33) \div (\log 30 - \log 10) = 0.321897$$

Calculator routine for K (enter divisor first):

Symbol	Enter	Press			Display
X_b	30	log	–		
X_a	10	log	=	Sto	
T_a	47	log	–		
T_b	33	log	=	÷	
		Rcl	=		0.321897 for K

2. The unit one amount (T_1) is calculated next.

$$T_1 = X^K \cdot T_a$$

Substituting the data in the formula:

$$T_1 = 30^{0.321987} \cdot 33 = 98.63 \text{ hr at unit one}$$

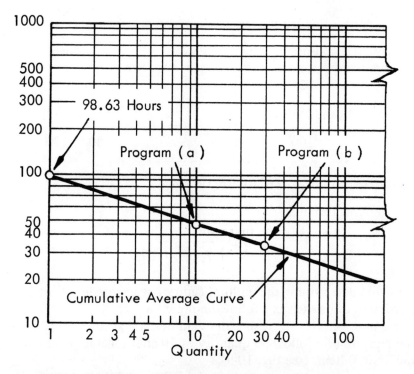

Fig. 10-8. Calculated cumulative average curve based on two data points.

Calculator routine for unit one:

Symbol	Enter	Press	Display
X	30	y^x	
K	0.321897	X	
T_A	33	=	98.63 hr at T_1

3. The learning-curve percent (P) is calculated by determining the learning-curve percent from the K value:

$$P = 2^{-K} \cdot 100$$

Substituting the data in the formula:

$$P = 2^{-0.321897} \cdot 100 = 80.0017\%$$

Calculator routine for the learning-curve percent:

Symbol	Enter	Press		Display
	2	y^x		
K	0.321897	$+/-$	$\times$	
	100	$=$		80.0017%

MIDUNIT--UNIT OF AVERAGE VALUE

The midunit formula:

$$X = \sqrt[K]{T_1 \cdot L \div T_U}$$

is used in determining the midunit (unit of average value) for a lot, or a group of units such as a follow-on program.

Example—problem: Using the data developed for Fig. 10-8, find the midunit of a follow-on quantity of 70 additional items, following the first 30 items (see Fig. 10-9).

The learning curve was rounded to 80% and the L and K values were taken from Fig. 10-4.

$$T_U = 17.86 \text{ hr average for 70 follow-on units}$$

Substituting the given values in the formula:

$$X = \frac{0.321928}{\sqrt{98.64 \cdot 0.678072 \div 17.86}}$$

Calculator routine:

Symbol	Enter	Press	Display
T_1	98.64	$\times$	
L	0.678072	$\div$	
T_U	17.86	$\sqrt[x]{y}$	
K	0.321928	$=$	60.43th unit—use 61st unit.

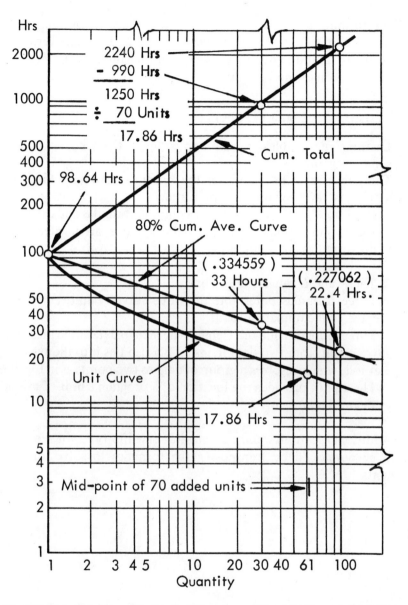

Fig. 10-9. Calculating the midpoint of a follow-on quantity, using the pocket calculator.

This formula is based on the T_U approximation formula, so the next higher unit (sixty-first) should be chosen. The same formula is applicable to an initial quantity, for which the cumulative average for the group becomes the T_U value.

CONCLUSIONS AND MAJOR POINTS

1. Learning-curve tables are ample to compute the needed values for some programs.
2. The pocket calculator reduces the learning-curve mathematics to a practical level for routine application.
3. A small notebook containing the common formulas and calculator routines, plus an inexpensive pocket calculator, are needed tools.
4. Eight formulas are required:

   The K and L values require one.

   Three formulas determine the three curves: unit, cumulative average, and cumulative total.

   Three more provide the elements needed to determine a learning-curve graph from two data items. This also ties into the averaging method, which resolves a group of data to two points as in Fig. 9-6.

   The eighth provides the location of a program's unit of average value on the unit curve.

Additional formulas and routines for specific types of work can often be developed by transposing the formulas to solve for other elements.

APPENDIX I - - PERIOD DISTRIBUTION CURVES

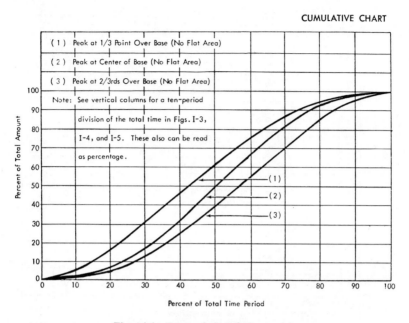

(1) Peak at 1/3 Point Over Base (No Flat Area)

(2) Peak at Center of Base (No Flat Area)

(3) Peak at 2/3rds Over Base (No Flat Area)

Note: See vertical columns for a ten-period division of the total time in Figs. I-3, I-4, and I-5. These also can be read as percentage.

Percent of Total Amount

Percent of Total Time Period

Fig. A1-1. Curves, 1, 2, & 3 Compared

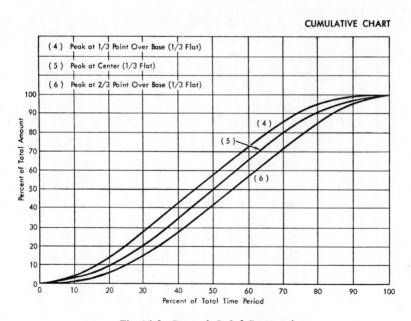

(4) Peak at 1/3 Point Over Base (1/3 Flat)

(5) Peak at Center (1/3 Flat)

(6) Peak at 2/3 Point Over Base (1/3 Flat)

Percent of Total Amount

Percent of Total Time Period

Fig. A1-2. Curves 4, 5, & 6 Compared

161

CURVE NO. 1.   Peak at 1/3rd, No Flat Area   Note: Columns are read vertically

PERCENTAGE PER PERIOD

Period	3	4	5	6	7	8	9	10	11	12	13	14	15	16	17	18	19	20	21	22	23	24	25
1	34.5	21.6	14.6	10.4	7.6	5.6	4.6	3.7	3.1	2.7	2.3	2.1	1.8	1.6	1.5	1.4	1.3	1.2	1.1	1.0	1.0	1.0	0.9
2	48.8	39.3	30.6	24.1	19.6	16.0	12.9	10.9	9.2	7.7	6.7	5.5	4.8	4.0	3.7	3.2	2.9	2.5	2.3	2.1	1.8	1.7	1.6
3	16.7	30.8	30.0	26.4	22.5	19.5	17.0	14.8	12.8	11.2	9.9	8.9	8.0	7.4	6.6	5.8	5.1	4.6	4.2	3.7	3.5	2.9	2.7
4		8.3	20.1	22.4	21.7	19.8	17.8	15.8	14.3	12.9	11.7	10.7	9.5	8.6	7.8	7.1	6.8	6.3	5.9	5.5	4.9	4.8	4.4
5			4.7	13.8	17.1	17.5	17.0	15.7	14.4	13.3	12.1	11.2	10.4	9.6	9.0	8.3	7.5	7.0	6.3	6.0	5.6	5.4	5.0
6				2.9	9.5	13.3	14.0	14.3	14.0	13.1	12.3	11.3	10.7	9.9	9.3	8.7	8.2	7.8	7.3	6.8	6.6	5.8	5.5
7					2.0	6.8	10.5	11.8	12.0	12.1	11.7	11.3	10.5	10.0	9.3	8.9	8.4	7.9	7.4	7.1	6.8	6.3	6.2
8						1.5	5.0	8.3	10.0	10.3	10.6	10.4	10.2	9.8	9.2	8.9	8.4	7.9	7.6	7.2	6.8	6.6	6.2
9							1.2	3.7	6.5	8.4	9.0	9.3	9.3	9.2	9.0	8.6	8.3	7.9	7.6	7.2	6.9	6.7	6.3
10								1.0	2.9	5.4	6.8	7.8	8.1	8.3	8.3	8.4	8.0	7.5	7.1	7.1	6.8	6.6	6.4
11									0.8	2.1	4.5	5.8	6.9	7.2	7.4	7.4	7.4	6.8	7.0	6.9	6.8	6.5	6.2
12										0.8	1.7	3.7	5.1	6.1	6.5	6.6	6.8	6.2	6.2	6.3	6.2	6.4	6.2
13											0.7	1.4	3.0	4.4	5.3	5.8	6.2	5.6	5.7	5.7	5.7	6.4	6.2
14												0.6	1.2	2.4	3.7	4.7	5.3	4.7	5.2	5.2	5.5	5.7	5.8
15													0.5	1.0	2.1	3.3	4.1	3.6	4.1	4.8	4.8	5.4	5.4
16														0.5	0.9	1.7	2.8	2.4	3.2	3.6	4.3	4.9	4.9
17															0.4	0.8	1.4	1.3	2.2	2.9	3.4	4.5	4.7
18																0.4	0.7	0.6	1.1	1.9	2.5	3.9	4.3
19																	0.4	0.4	0.5	1.0	1.7	3.0	3.4
20																			0.4	0.5	0.8	2.4	2.8
21																				0.3	0.5	1.4	2.0
22																					0.3	0.7	1.3
23																						0.5	0.7
24																						0.3	0.4
25																							0.3

1/3   2/3

CURVE NO. 2. Peak at Center, No Flat Area   Note: Columns are read vertically

PERCENTAGE PER PERIOD

Period	3	4	5	6	7	8	9	10	11	12	13	14	15	16	17	18	19	20	21	22	23	24	25
1	20.8	10.3	6.1	4.0	2.7	2.1	1.6	1.2	1.0	0.9	0.8	0.7	0.7	0.6	0.6	0.5	0.5	0.4	0.4	0.3	0.3	0.3	0.3
2	58.4	39.7	26.0	16.8	11.4	8.2	6.3	4.9	3.5	3.1	2.5	2.0	1.6	1.5	1.3	1.1	1.0	0.8	0.8	0.7	0.7	0.6	0.6
3	20.8	39.7	35.8	29.2	23.0	17.3	12.9	9.9	8.2	6.3	5.3	4.5	3.8	3.1	2.7	2.4	2.0	1.9	1.5	1.3	1.3	1.2	1.0
4		10.3	26.0	29.2	25.8	22.4	19.1	16.1	13.4	10.5	8.5	6.9	6.1	5.1	4.5	3.9	3.4	3.0	2.5	2.2	2.1	1.9	1.7
5			6.1	16.8	23.0	22.4	20.2	17.9	15.8	14.1	12.0	10.5	8.6	7.5	6.4	5.6	4.9	4.2	3.7	3.4	3.0	2.7	2.5
6				4.0	11.4	17.3	19.1	17.9	16.2	15.1	13.9	12.5	11.3	9.8	8.6	7.3	6.4	5.7	5.2	4.8	4.0	3.6	3.4
7					2.7	8.2	12.9	16.1	15.8	15.1	14.0	12.9	11.9	11.1	10.1	9.2	8.1	7.5	6.7	6.2	5.1	4.7	4.3
8						2.1	6.3	9.9	13.4	14.1	13.9	12.9	12.0	11.3	10.5	9.9	9.4	8.6	7.8	7.2	6.2	5.8	5.3
9							1.6	4.9	8.2	10.5	12.0	12.5	11.9	11.3	10.6	10.1	9.5	9.0	8.5	7.8	7.2	6.8	6.2
10								1.2	3.5	6.3	8.5	10.5	11.3	11.1	10.5	10.1	9.6	9.0	8.6	8.0	7.9	7.3	6.8
11									1.0	3.1	5.3	6.9	8.6	9.8	10.1	9.9	9.5	9.0	8.6	8.1	8.1	7.5	7.1
12										0.9	2.5	4.5	6.1	7.5	8.6	9.2	9.4	8.9	8.6	8.1	8.2	7.6	7.2
13											0.8	2.0	3.8	5.1	6.4	7.3	8.1	8.6	8.5	8.0	8.1	7.6	7.2
14												0.7	1.6	3.1	4.5	5.6	6.4	7.5	7.8	7.8	7.9	7.5	7.2
15													0.7	1.5	2.7	3.9	4.9	5.7	6.7	7.2	7.2	7.3	7.1
16														0.6	1.3	2.4	3.4	4.2	5.2	6.2	6.2	6.8	6.8
17															0.6	1.1	2.0	3.0	3.7	4.8	5.1	5.8	6.2
18																0.5	1.0	1.9	2.5	3.4	4.0	4.7	5.3
19																	0.5	0.8	1.5	2.2	3.0	3.6	4.3
20																		0.4	0.8	1.3	2.1	2.7	3.4
21																			0.4	0.7	1.3	1.9	2.5
22																				0.3	0.7	1.2	1.7
23																					0.3	0.6	1.0
24																						0.3	0.6
25																							0.3

CURVE NO. 3. Peak at 2/3rds, No Flat Area   Note: Columns are read vertically

PERCENT PER PERIOD

Period	3	4	5	6	7	8	9	10	11	12	13	14	15	16	17	18	19	20	21	22	23	24	25
1	16.7	8.3	4.7	2.9	2.0	1.5	1.2	1.0	0.8	0.8	0.7	0.6	0.5	0.5	0.4	0.4	0.4	0.4	0.4	0.3	0.3	0.3	0.3
2	48.8	30.8	20.1	13.8	9.5	6.8	5.0	3.7	2.9	2.1	1.7	1.4	1.2	1.0	0.9	0.8	0.7	0.6	0.5	0.5	0.5	0.5	0.4
3	34.5	39.3	30.0	22.4	17.1	13.3	10.5	8.3	6.5	5.4	4.5	3.7	3.0	2.4	2.1	1.7	1.4	1.3	1.1	1.0	0.8	0.7	0.7
4		21.6	30.6	26.4	21.7	17.5	14.0	11.8	10.0	8.4	6.8	5.8	5.1	4.4	3.7	3.3	2.8	2.4	2.2	1.9	1.7	1.4	1.3
5			14.6	24.1	22.5	19.8	17.0	14.3	12.0	10.3	9.0	7.8	6.9	6.1	5.3	4.7	4.1	3.6	3.2	2.9	2.5	2.4	2.0
6				10.4	19.6	19.5	17.8	15.7	14.0	12.1	10.6	9.3	8.1	7.2	6.5	5.8	5.3	4.7	4.1	3.6	3.4	3.0	2.8
7					7.6	16.0	17.0	15.8	14.4	13.1	11.7	10.4	9.3	8.3	7.4	6.6	6.2	5.6	5.2	4.8	4.3	3.9	3.4
8						5.6	12.9	14.8	14.3	13.3	12.3	11.3	10.2	9.2	8.3	7.4	6.3	6.2	5.7	5.2	4.8	4.5	4.3
9							4.6	10.9	12.8	12.9	12.1	11.3	10.5	9.8	9.0	8.4	7.4	6.8	6.2	5.7	5.5	4.9	4.7
10								3.7	9.2	11.2	11.7	11.2	10.7	10.0	9.2	8.6	8.0	7.5	7.0	6.3	5.7	5.4	4.9
11									3.1	7.7	9.9	10.7	10.4	9.9	9.3	8.9	8.3	7.9	7.6	6.9	6.2	5.7	5.4
12										2.7	6.7	8.9	9.5	9.6	9.3	8.9	8.4	7.9	7.6	7.1	6.8	6.4	5.8
13											2.3	5.5	8.0	8.6	9.0	8.7	8.4	7.9	7.6	7.2	6.8	6.5	6.2
14												2.1	4.8	7.4	7.8	8.3	8.2	7.8	7.4	7.2	6.8	6.6	6.2
15													1.3	4.0	6.6	7.1	7.5	7.0	7.3	7.2	6.9	6.7	6.4
16														1.6	3.7	5.8	6.8	6.3	6.3	7.1	6.8	6.6	6.3
17															1.5	3.2	5.1	4.6	5.9	6.8	6.8	6.6	6.2
18																1.4	2.9	2.5	4.2	6.0	5.6	6.3	6.2
19																	1.3	1.2	2.3	5.5	4.9	5.8	5.5
20																			1.1	3.7	3.5	5.4	5.0
21																				2.1	1.8	4.8	4.4
22																				1.0	1.0	2.9	2.7
23																						1.7	1.6
24																						1.0	0.9
25																							

CURVE NO. 4 . Peak at 1/3rd, One Third Flat   Note: Columns are read vertically

PERCENTAGE PER PERIOD

Period	3	4	5	6	7	8	9	10	11	12	13	14	15	16	17	18	19	20	21	22	23	24	25
1	33.2	21.1	14.2	10.0	7.0	5.1	4.0	3.0	2.5	2.1	1.7	1.5	1.2	1.1	0.9	0.9	0.8	0.8	0.7	0.6	0.6	0.6	0.5
2	47.8	36.5	28.7	23.2	19.3	16.0	13.3	11.2	9.4	7.9	6.6	5.5	4.8	4.0	3.5	3.1	2.7	2.2	2.0	1.9	1.6	1.5	1.3
3	19.0	33.0	29.2	24.4	20.8	18.2	15.9	14.1	12.5	11.1	10.2	9.2	8.2	7.5	6.9	6.0	5.3	4.8	4.3	3.8	3.3	3.0	2.9
4		9.4	22.6	23.4	20.8	18.3	16.3	14.6	13.2	12.1	11.1	10.1	9.3	8.5	7.8	7.3	7.0	6.4	6.1	5.6	5.2	4.9	4.3
5			5.3	15.6	19.0	18.0	16.2	14.7	13.4	12.2	11.2	10.3	9.7	9.1	8.5	7.9	7.4	6.9	6.4	6.0	5.9	5.4	5.2
6				3.4	10.8	15.0	15.3	14.5	13.3	12.2	11.2	10.5	9.7	9.1	8.6	8.0	7.5	7.2	6.8	6.5	6.3	5.7	5.6
7					2.3	7.8	12.0	13.0	13.0	12.2	11.2	10.5	9.8	9.2	8.6	8.1	7.7	7.3	6.9	6.6	6.4	6.0	5.7
8						1.6	5.7	9.6	11.0	11.2	11.1	10.3	9.7	9.1	8.6	8.2	7.7	7.3	6.9	6.6	6.4	6.1	5.8
9							1.3	4.3	7.4	9.6	10.0	10.0	9.7	9.1	8.5	8.1	7.7	7.4	7.0	6.7	6.4	6.1	5.8
10								1.0	3.4	6.0	8.1	9.0	8.9	8.9	8.5	8.0	7.7	7.3	6.9	6.7	6.4	6.1	5.8
11									0.9	2.7	4.9	6.8	7.8	8.1	8.2	8.0	7.4	7.3	6.9	6.7	6.3	6.1	5.9
12										0.7	2.1	4.0	5.9	6.9	7.3	7.4	6.8	7.2	6.7	6.6	6.3	6.1	5.9
13											0.6	1.7	3.4	4.9	6.0	6.7	5.8	6.8	6.4	6.6	6.2	6.1	5.9
14												0.6	1.4	2.9	4.2	5.3	4.7	6.2	5.9	6.4	6.1	5.8	5.8
15													0.5	1.2	2.5	3.6	3.2	5.5	4.8	5.9	5.9	5.4	5.8
16														0.4	1.0	2.1	1.8	4.1	3.6	5.1	5.4	5.0	5.5
17															0.4	0.9	0.8	2.7	2.4	4.2	4.8	4.6	5.1
18																0.4	0.3	1.6	1.4	3.2	3.9	3.5	4.7
19																	0.3	0.7	0.6	2.2	2.8	2.5	4.1
20																		0.3	0.3	1.2	1.9	1.8	3.2
21																				0.6	1.0	0.9	2.3
22																				0.3	0.5	0.5	1.5
23																					0.3	0.2	0.8
24																							0.5
25																							0.2

1/3   2/3

CURVE NO. 5.   Peak at Center, 1/3 Flat   Note: Columns are read vertically

PERCENTAGE PER PERIOD

Period	3	4	5	6	7	8	9	10	11	12	13	14	15	16	17	18	19	20	21	22	23	24	25
1	26.1	14.8	8.9	5.6	4.0	2.9	2.2	2.0	1.7	1.4	1.3	1.2	1.1	1.0	0.9	0.8	0.7	0.7	0.6	0.6	0.6	0.5	0.5
2	47.8	35.2	26.6	20.5	15.5	11.9	9.2	6.9	5.3	4.2	3.3	2.8	2.2	1.9	1.7	1.4	1.4	1.3	1.2	1.1	1.0	0.9	0.8
3	26.1	35.2	29.0	23.9	20.3	17.2	14.7	12.6	10.7	9.2	7.9	6.5	5.6	4.6	3.9	3.4	2.7	2.4	2.2	1.8	1.6	1.5	1.4
4		14.8	26.6	23.9	20.4	18.0	15.9	14.0	12.6	11.3	10.0	9.0	7.9	7.3	6.5	5.8	5.3	4.5	4.0	3.5	3.1	2.7	2.4
5			8.9	20.5	20.3	18.0	16.0	14.5	13.1	11.9	11.0	10.1	9.3	8.4	7.6	7.0	6.3	5.9	5.3	5.0	4.5	4.2	3.8
6				5.6	15.5	17.2	15.9	14.5	13.2	12.0	11.0	10.2	9.5	8.8	8.1	7.7	7.1	6.6	6.2	5.7	5.3	5.0	4.6
7					4.0	11.9	14.7	14.0	13.1	12.0	11.0	10.2	9.6	9.0	8.4	7.9	7.5	6.9	6.6	6.2	5.8	5.5	5.2
8						2.9	9.2	12.6	13.1	11.9	11.0	10.2	9.6	9.0	8.6	8.0	7.6	7.2	6.8	6.4	6.1	5.8	5.5
9							2.2	6.9	10.7	11.3	11.0	10.2	9.6	9.0	8.6	8.0	7.6	7.2	6.8	6.5	6.2	5.9	5.6
10								2.0	5.3	9.2	10.0	10.1	9.5	9.0	8.6	8.0	7.6	7.3	6.8	6.6	6.3	6.0	5.7
11									1.7	4.2	7.9	9.0	9.3	8.8	8.4	8.0	7.6	7.3	6.8	6.6	6.3	6.0	5.8
12										1.4	3.3	6.5	7.9	8.4	8.1	7.9	7.6	7.2	6.8	6.6	6.3	6.0	5.8
13											1.3	2.8	5.6	7.3	7.6	7.7	7.5	7.2	6.8	6.6	6.3	6.0	5.8
14												1.2	2.2	4.6	6.5	7.0	7.1	6.9	6.8	6.5	6.3	6.0	5.8
15													1.1	1.9	3.9	5.8	6.3	6.6	6.6	6.4	6.2	6.0	5.8
16														1.0	1.7	3.4	5.3	5.9	6.2	6.2	6.1	5.9	5.7
17															0.9	1.4	2.7	4.5	5.3	5.7	5.8	5.8	5.6
18																0.8	1.4	2.4	4.0	5.0	5.3	5.5	5.5
19																	0.7	1.3	2.2	3.5	4.5	5.0	5.2
20																		0.7	1.2	1.8	3.1	4.2	4.6
21																			0.6	1.1	1.6	2.7	3.8
22																				0.6	1.0	1.5	2.4
23																					0.6	0.9	1.4
24																						0.5	0.8
25																							0.5

CURVE NO. 6. Peak at 2/3rds, One Third Flat   Note: Columns are read vertically

PERCENTAGE PER PERIOD

Period	3	4	5	6	7	8	9	10	11	12	13	14	15	16	17	18	19	20	21	22	23	24	25
1	19.0	9.4	5.3	3.4	2.3	1.6	1.3	1.0	0.9	0.7	0.6	0.6	0.5	0.4	0.4	0.4	0.3	0.3	0.3	0.3	0.3	0.2	0.2
2	47.8	33.0	22.6	15.6	10.8	7.8	5.7	4.3	3.4	2.7	2.1	1.7	1.4	1.2	1.0	0.9	0.8	0.7	0.6	0.6	0.5	0.5	0.5
3	33.2	36.5	29.2	23.4	19.0	15.0	12.0	9.6	7.4	6.0	4.9	4.0	3.4	2.9	2.5	2.1	1.8	1.6	1.4	1.2	1.0	0.9	0.8
4		21.1	28.7	24.4	20.8	18.0	15.3	13.0	11.0	9.6	8.1	6.8	5.9	4.9	4.2	3.6	3.2	2.7	2.4	2.2	1.9	1.8	1.5
5			14.2	23.2	20.8	18.3	16.2	14.5	13.0	11.2	10.0	9.0	7.8	6.9	6.0	5.3	4.7	4.1	3.6	3.2	2.8	2.5	2.3
6				10.0	19.3	18.2	16.3	14.7	13.3	12.2	11.1	10.0	8.9	8.1	7.3	6.7	5.8	5.5	4.8	4.2	3.9	3.5	3.2
7					7.0	16.0	15.9	14.6	13.4	12.2	11.2	10.3	9.7	8.9	8.2	7.4	6.8	6.2	5.9	5.1	4.8	4.6	4.1
8						5.1	13.3	14.1	13.2	12.1	11.2	10.5	9.7	9.1	8.5	8.0	7.4	6.8	6.4	5.9	5.4	5.0	4.7
9							4.0	11.2	12.5	12.1	11.1	10.5	9.8	9.1	8.5	8.0	7.7	7.2	6.7	6.4	5.9	5.4	5.1
10								3.0	9.4	11.1	11.1	10.3	9.7	9.2	8.6	8.1	7.7	7.3	6.9	6.6	6.1	5.8	5.5
11									2.5	7.9	10.2	10.1	9.7	9.1	8.6	8.2	7.7	7.3	6.9	6.6	6.2	6.1	5.8
12										2.1	6.6	9.2	9.3	9.1	8.6	8.1	7.7	7.4	7.0	6.7	6.3	6.1	5.8
13											1.7	5.5	8.2	8.5	8.5	8.0	7.7	7.3	7.0	6.7	6.4	6.1	5.9
14												1.5	4.8	7.5	7.8	7.9	7.5	7.3	6.9	6.7	6.4	6.1	5.9
15													1.2	4.0	6.9	7.3	7.4	7.2	6.9	6.6	6.4	6.1	5.8
16														1.1	3.5	6.0	7.0	6.9	6.8	6.6	6.4	6.1	5.8
17															0.9	3.1	5.3	6.4	6.4	6.5	6.4	6.1	5.8
18																0.9	2.7	4.8	6.1	6.0	6.3	6.0	5.8
19																	0.8	2.2	4.3	5.6	5.9	5.7	5.7
20																		0.8	2.0	3.8	5.2	5.4	5.6
21																			0.7	1.9	3.3	4.9	5.2
22																				0.6	1.6	3.0	4.3
23																					0.6	1.5	2.9
24																						0.6	1.3
25																							0.5

2/3

1/3

APPENDIX 2 - LEARNING-CURVE TABLES

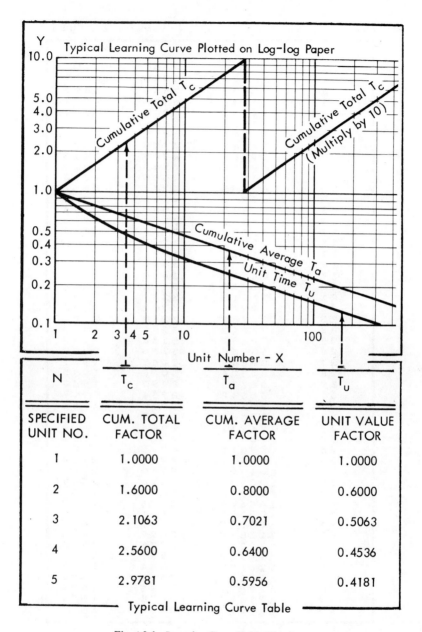

N	T_c	T_a	T_u
SPECIFIED UNIT NO.	CUM. TOTAL FACTOR	CUM. AVERAGE FACTOR	UNIT VALUE FACTOR
1	1.0000	1.0000	1.0000
2	1.6000	0.8000	0.6000
3	2.1063	0.7021	0.5063
4	2.5600	0.6400	0.4536
5	2.9781	0.5956	0.4181

Typical Learning Curve Table

Fig. A2-1. Learning-Curve Table Diagram

78%   LEARNING-CURVE TABLES

UNIT	T C	T A	T U
1	10000	10000	10000
2	15600	7800	5600
3	20234	6744	4634
4	24336	6084	4101
5	28081	5616	3745
6	31566	5261	3484
7	34847	4978	3281
8	37964	4745	3116
9	40944	4549	2979
10	43807	4380	2863
11	46569	4233	2762
12	49242	4103	2673
13	51837	3987	2594
14	54361	3882	2524
15	56822	3788	2460
16	59224	3701	2402
17	61572	3621	2348
18	63872	3548	2299
19	66127	3480	2254
20	68339	3416	2212
21	70512	3357	2172
22	72648	3302	2136
23	74749	3249	2101
24	76819	3200	2069
25	78857	3154	2038
26	80866	3110	2009
27	82848	3068	1981
28	84804	3028	1955
29	86735	2990	1930
30	88642	2954	1907
31	90526	2920	1884
32	92389	2887	1862
33	94231	2855	1842
34	96053	2825	1822
35	97856	2795	1803
36	99641	2767	1784
37	101408	2740	1766
38	103158	2714	1749
39	104891	2689	1733
40	106609	2665	1717
41	108311	2641	1702
42	109999	2619	1687
43	111672	2597	1673
44	113331	2575	1659
45	114977	2555	1645
46	116609	2534	1632
47	118230	2515	1620
48	119837	2496	1607
49	121433	2478	1595
50	123017	2460	1584

78%   1 - 50

Appendix II

LEARNING-CURVE TABLES   78%

UNIT	T C	T A	T U
51	124590	2442	1572
52	126152	2426	1561
53	127703	2409	1551
34	129243	2393	1540
55	130774	2377	1530
56	132294	2362	1520
57	133805	2347	1510
58	135307	2332	1501
59	136799	2318	1492
60	138282	2304	1483
61	139756	2291	1474
62	141221	2277	1465
63	142678	2264	1457
64	144127	2251	1448
65	145568	2239	1440
66	147001	2227	1432
67	148426	2215	1425
68	149843	2203	1417
69	151253	2192	1410
70	152656	2180	1402
71	154052	2169	1395
72	155440	2158	1388
73	156822	2148	1381
74	158196	2137	1374
75	159565	2127	1368
76	160926	2117	1361
77	162282	2107	1355
78	163631	2097	1348
79	164973	2088	1342
80	166310	2078	1336
81	167641	2069	1330
82	168966	2060	1324
83	170285	2051	1319
84	171598	2042	1313
85	172906	2034	1307
86	174208	2025	1302
87	175505	2017	1296
88	176797	2009	1291
89	178083	2000	1286
90	179364	1992	1281
91	180640	1985	1276
92	181911	1977	1271
93	183177	1969	1266
94	184438	1962	1261
95	185695	1954	1256
96	186946	1947	1251
97	188193	1940	1247
98	189436	1933	1242
99	190674	1926	1237
100	191907	1919	1233

79%   LEARNING-CURVE TABLES

UNIT	T_C	T_A	T_U
1	10000	10000	10000
2	15800	7900	5800
3	20647	6882	4847
4	24964	6241	4316
5	28924	5784	3960
6	32622	5437	3698
7	36116	5159	3493
8	39443	4930	3327
9	42631	4736	3188
10	45700	4570	3069
11	48667	4424	2966
12	51544	4295	2876
13	54339	4179	2795
14	57063	4075	2723
15	59721	3981	2658
16	62320	3895	2598
17	64863	3815	2543
18	67357	3742	2493
19	69804	3673	2446
20	72207	3610	2403
21	74570	3550	2362
22	76894	3495	2324
23	79184	3442	2289
24	81439	3393	2255
25	83663	3346	2223
26	85857	3302	2193
27	88022	3260	2165
28	90160	3220	2138
29	92272	3181	2112
30	94360	3145	2087
31	96424	3110	2064
32	98465	3077	2041
33	100485	3045	2019
34	102485	3014	1999
35	104464	2984	1979
36	106424	2956	1960
37	108366	2928	1941
38	110290	2902	1924
39	112197	2876	1906
40	114087	2852	1890
41	115962	2828	1874
42	117820	2805	1858
43	119664	2782	1843
44	121494	2761	1829
45	123309	2740	1815
46	125110	2719	1801
47	126899	2699	1788
48	128674	2680	1775
49	130437	2661	1762
50	132188	2643	1750

79%   1 - 50

Appendix II

LEARNING-CURVE TABLES   79%

UNIT	T C	T A	T U
51	133926	2626	1738
52	135654	2608	1727
53	137370	2591	1715
54	139075	2575	1705
55	140769	2559	1694
56	142453	2543	1683
57	144126	2528	1673
58	145790	2513	1663
59	147444	2499	1653
60	149089	2484	1644
61	150724	2470	1635
62	152350	2457	1626
63	153967	2443	1617
64	155576	2430	1608
65	157175	2418	1599
66	158767	2405	1591
67	160351	2393	1583
68	161926	2381	1575
69	163493	2369	1567
70	165053	2357	1559
71	166606	2346	1552
72	168150	2335	1544
73	169688	2324	1537
74	171219	2313	1530
75	172742	2303	1523
76	174258	2292	1516
77	175768	2282	1509
78	177271	2272	1503
79	178768	2262	1496
80	180258	2253	1490
81	181742	2243	1483
82	183219	2234	1477
83	184691	2225	1471
84	186156	2216	1465
85	187616	2207	1459
86	189070	2198	1453
87	190518	2189	1447
88	191960	2181	1442
89	193397	2173	1436
90	194828	2164	1431
91	196254	2156	1425
92	197675	2148	1420
93	199090	2140	1415
94	200500	2132	1410
95	201905	2125	1405
96	203305	2117	1400
97	204700	2110	1395
98	206090	2102	1390
99	207476	2095	1385
100	208857	2088	1380

79%   51 - 100
                                       Appendix II

80% LEARNING-CURVE TABLES

UNIT	T C	T A	T U
1	10000	10000	10000
2	16000	8000	6000
3	21063	7021	5063
4	25600	6400	4536
5	29781	5956	4181
6	33700	5616	3919
7	37414	5344	3713
8	40960	5120	3545
9	44365	4929	3405
10	47650	4765	3285
11	50832	4621	3181
12	53921	4493	3089
13	56929	4379	3007
14	59862	4275	2933
15	62729	4181	2867
16	65536	4096	2806
17	68286	4016	2750
18	70984	3943	2698
19	73635	3875	2650
20	76241	3812	2606
21	78806	3752	2564
22	81331	3696	2525
23	83820	3644	2488
24	86274	3594	2454
25	88695	3547	2421
26	91086	3503	2390
27	93447	3461	2361
28	95780	3420	2333
29	98086	3382	2306
30	100367	3345	2280
31	102624	3310	2256
32	104857	3276	2233
33	107068	3244	2210
34	109257	3213	2189
35	111426	3183	2168
36	113575	3154	2148
37	115705	3127	2129
38	117816	3100	2111
39	119910	3074	2093
40	121986	3049	2076
41	124046	3025	2059
42	126089	3002	2043
43	128117	2979	2027
44	130130	2957	2012
45	132128	2936	1998
46	134112	2915	1983
47	136082	2895	1970
48	138039	2875	1956
49	139982	2856	1943
50	141913	2838	1930

Appendix II

LEARNING-CURVE TABLES   80%

UNIT	T C	T A	T U
51	143831	2820	1918
52	145738	2802	1906
53	147632	2785	1894
54	149516	2768	1883
55	151387	2752	1871
56	153248	2736	1860
57	155099	2721	1850
58	156939	2705	1839
59	158768	2690	1829
60	160588	2676	1819
61	162398	2662	1810
62	164199	2648	1800
63	165990	2634	1791
64	167772	2621	1782
65	169545	2608	1773
66	171309	2595	1764
67	173065	2583	1755
68	174812	2570	1747
69	176551	2558	1739
70	178282	2546	1730
71	180005	2535	1723
72	181720	2523	1715
73	183428	2512	1707
74	185128	2501	1700
75	186821	2490	1692
76	188506	2480	1685
77	190185	2469	1678
78	191856	2459	1671
79	193520	2449	1664
80	195178	2439	1657
81	196829	2429	1651
82	198473	2420	1644
83	200111	2410	1638
84	201743	2401	1631
85	203369	2392	1625
86	204988	2383	1619
87	206601	2374	1613
88	208208	2366	1607
89	209810	2357	1601
90	211405	2348	1595
91	212995	2340	1589
92	214580	2332	1584
93	216158	2324	1578
94	217732	2316	1573
95	219300	2308	1567
96	220862	2300	1562
97	222420	2292	1557
98	223972	2285	1552
99	225519	2277	1547
100	227061	2270	1542

81%   LEARNING-CURVE TABLES

UNIT	T C	T A	T U
1	10000	10000	10000
2	16200	8100	6200
3	21481	7160	5281
4	26244	6561	4762
5	30653	6130	4409
6	34800	5800	4147
7	38742	5534	3941
8	42515	5314	3773
9	46147	5127	3632
10	49658	4965	3511
11	53064	4824	3405
12	56377	4698	3312
13	59607	4585	3229
14	62762	4483	3155
15	65849	4389	3087
16	68874	4304	3025
17	71843	4226	2968
18	74758	4153	2915
19	77625	4085	2866
20	80446	4022	2821
21	83225	3963	2778
22	85964	3907	2738
23	88665	3855	2701
24	91331	3805	2665
25	93963	3758	2632
26	96563	3713	2600
27	99133	3671	2570
28	101674	3631	2541
29	104188	3592	2513
30	106676	3555	2487
31	109138	3520	2462
32	111577	3486	2438
33	113992	3454	2415
34	116385	3423	2393
35	118757	3393	2371
36	121109	3364	2351
37	123440	3336	2331
38	125753	3309	2312
39	128047	3283	2294
40	130323	3258	2276
41	132582	3233	2259
42	134825	3210	2242
43	137051	3187	2226
44	139262	3165	2210
45	141457	3143	2195
46	143637	3122	2180
47	145804	3102	2166
48	147956	3082	2152
49	150094	3063	2138
50	152220	3044	2125

81%   1 - 50

Appendix II

LEARNING-CURVE TABLES 81%

UNIT	T C	T A	T U
51	154332	3026	2112
52	156432	3008	2099
53	158520	2990	2087
54	160596	2974	2075
55	162660	2957	2064
56	164713	2941	2052
57	166754	2925	2041
58	168785	2910	2030
59	170805	2895	2020
60	172815	2880	2009
61	174814	2865	1999
62	176804	2851	1989
63	178784	2837	1979
64	180754	2824	1970
65	182715	2811	1961
66	184667	2797	1951
67	186610	2785	1942
68	188544	2772	1934
69	190470	2760	1925
70	192387	2748	1917
71	194296	2736	1908
72	196196	2724	1900
73	198089	2713	1892
74	199974	2702	1884
75	201850	2691	1876
76	203720	2680	1869
77	205582	2669	1861
78	207436	2659	1854
79	209284	2649	1847
80	211124	2639	1840
81	212957	2629	1833
82	214784	2619	1826
83	216603	2609	1819
84	218416	2600	1813
85	220223	2590	1806
86	222023	2581	1800
87	223817	2572	1793
88	225604	2563	1787
89	227385	2554	1781
90	229160	2546	1775
91	230930	2537	1769
92	232693	2529	1763
93	234450	2520	1757
94	236202	2512	1751
95	237948	2504	1746
96	239689	2496	1740
97	241424	2488	1734
98	243153	2481	1729
99	244877	2473	1724
100	246596	2465	1718

81% 51 - 100

Appendix II

82%   LEARNING-CURVE TABLES

UNIT	T_C	T_A	T_U
1	10000	10000	10000
2	16400	8200	6400
3	21903	7301	5503
4	26896	6724	4992
5	31539	6807	4643
6	35922	5987	4382
7	40099	5728	4177
8	44109	5513	4009
9	47977	5330	3868
10	51724	5172	3746
11	55365	5033	3640
12	58912	4909	3547
13	62375	4798	3463
14	65763	4697	3387
15	69083	4605	3319
16	72339	4521	3256
17	75538	4443	3198
18	78683	4371	3145
19	81778	4304	3095
20	84828	4241	3049
21	87834	4182	3005
22	90799	4127	2965
23	93725	4075	2926
24	96616	4025	2890
25	99472	3978	2856
26	102296	3934	2823
27	105089	3892	2792
28	107852	3851	2763
29	110587	3813	2735
30	113296	3776	2708
31	115978	3741	2682
32	118636	3707	2657
33	121271	3674	2634
34	123882	3643	2611
35	126472	3613	2589
36	129040	3584	2568
37	131588	3556	2548
38	134117	3529	2528
39	136626	3503	2509
40	139118	3477	2491
41	141591	3453	2473
42	144047	3429	2456
43	146487	3406	2439
44	148910	3384	2423
45	151318	3362	2407
46	153710	3341	2392
47	156088	3321	2377
48	158451	3301	2363
49	160800	3281	2348
50	163135	3262	2335

82%   1 - 50

Appendix II

LEARNING-CURVE TABLES   82%

UNIT	T C	T A	T U
51	165457	3244	2321
52	167766	3226	2308
53	170062	3208	2296
54	172346	3191	2283
55	174616	3174	2271
56	176878	3158	2260
57	179126	3142	2248
58	181364	3126	2237
59	183590	3111	2226
60	185805	3096	2215
61	188010	3082	2204
62	190205	3067	2194
63	192389	3053	2184
64	194564	3040	2174
65	196729	3026	2164
66	198884	3013	2155
67	201030	3000	2146
68	203167	2987	2136
69	205295	2975	2127
70	207414	2963	2119
71	209524	2951	2110
72	211626	2939	2101
73	213720	2927	2093
74	215805	2916	2085
75	217883	2905	2077
76	219952	2894	2069
77	222014	2883	2061
78	224068	2872	2054
79	226114	2862	2046
80	228153	2851	2039
81	230185	2841	2031
82	232210	2831	2024
83	234227	2822	2017
84	236238	2812	2010
85	238242	2802	2003
86	240239	2793	1997
87	242229	2784	1990
88	244213	2775	1983
89	246190	2766	1977
90	248161	2757	1971
91	250126	2748	1964
92	252085	2740	1958
93	254037	2731	1952
94	255984	2723	1946
95	257924	2714	1940
96	259859	2706	1934
97	261788	2698	1929
98	263712	2690	1923
99	265629	2683	1917
100	267541	2675	1912

82%   51 - 100
                                    Appendix II

83%   LEARNING-CURVE TABLES

UNIT	T_C	T_A	T_U
1	10000	10000	10000
2	16600	8300	6600
3	22328	7442	5728
4	27556	6889	5227
5	32439	6487	4883
6	37065	6177	4626
7	41487	5926	4422
8	45742	5717	4255
9	49857	5539	4114
10	53849	5384	3992
11	57736	5248	3886
12	61528	5127	3792
13	65237	5018	3708
14	68869	4919	3632
15	72433	4828	3563
16	75933	4745	3500
17	79374	4669	3441
18	82762	4597	3387
19	86100	4531	3337
20	89390	4469	3290
21	92637	4411	3246
22	95842	4356	3205
23	99008	4304	3166
24	102138	4255	3129
25	105232	4209	3094
26	108294	4165	3061
27	111324	4123	3030
28	114324	4083	2999
29	117295	4044	2971
30	120239	4007	2943
31	123156	3972	2917
32	126049	3939	2892
33	128917	3906	2868
34	131762	3875	2844
35	134585	3845	2822
36	137385	3816	2800
37	140166	3788	2780
38	142926	3761	2759
39	145666	3735	2740
40	148388	3709	2721
41	151091	3685	2703
42	153777	3661	2685
43	156446	3638	2668
44	159098	3615	2652
45	161734	3594	2635
46	164354	3572	2620
47	166959	3552	2604
48	169549	3532	2590
49	172124	3512	2575
50	174686	3493	2561

83%   1 - 50

Appendix II

LEARNING-CURVE TABLES   83 %

UNIT	T C	T A	T U
51	177233	3475	2547
52	179768	3457	2534
53	182289	3439	2521
54	184798	3422	2508
55	187294	3405	2496
56	189777	3388	2483
57	192249	3372	2472
58	194710	3357	2460
59	197159	3341	2448
60	199597	3326	2437
61	202024	3311	2426
62	204440	3297	2416
63	206846	3283	2405
64	209241	3269	2395
65	211627	3255	2385
66	214003	3242	2375
67	216369	3229	2366
68	218725	3216	2356
69	221072	3203	2347
70	223411	3191	2338
71	225740	3179	2329
72	228060	3167	2320
73	230372	3155	2311
74	232675	3144	2303
75	234970	3132	2294
76	237257	3121	2286
77	239535	3110	2278
78	241806	3100	2270
79	244069	3089	2262
80	246324	3079	2255
81	248572	3068	2247
82	250812	3058	2240
83	253045	3048	2232
84	255270	3038	2225
85	257489	3029	2218
86	259700	3019	2211
87	261905	3010	2204
88	264102	3001	2197
89	266293	2992	2191
90	268478	2983	2184
91	270656	2974	2177
92	272827	2965	2171
93	274993	2956	2165
94	277152	2948	2158
95	279304	2940	2152
96	281451	2931	2146
97	283592	2923	2140
98	285726	2915	2134
99	287855	2907	2128
100	289979	2899	2123

84 %   LEARNING-CURVE TABLES

UNIT	T C	T A	T U
1	10000	10000	10000
2	16800	8400	6800
3	22756	7585	5956
4	28224	7056	5467
5	33354	6670	5130
6	38231	6371	4876
7	42906	6129	4675
8	47416	5927	4509
9	51786	5754	4369
10	56035	5603	4249
11	60178	5470	4143
12	64228	5352	4049
13	68193	5245	3965
14	72082	5148	3889
15	75902	5060	3820
16	79659	4978	3756
17	83357	4903	3697
18	87000	4833	3643
19	90593	4768	3592
20	94139	4706	3545
21	97640	4649	3501
22	101100	4595	3459
23	104520	4544	3420
24	107903	4495	3383
25	111251	4450	3347
26	114565	4406	3314
27	117847	4364	3282
28	121099	4324	3251
29	124322	4286	3222
30	127517	4250	3194
31	130685	4215	3168
32	133827	4182	3142
33	136945	4149	3118
34	140040	4118	3094
35	143111	4088	3071
36	146161	4060	3049
37	149189	4032	3028
38	152197	4005	3007
39	155185	3979	2987
40	158153	3953	2968
41	161103	3929	2950
42	164036	3905	2932
43	166950	3882	2914
44	169848	3860	2897
45	172729	3838	2881
46	175594	3817	2864
47	178443	3796	2849
48	181277	3776	2834
49	184096	3757	2819
50	186901	3738	2804

LEARNING-CURVE TABLES   84 %

UNIT	T c	T A	T u
51	189692	3719	2790
52	192469	3701	2777
53	195233	3683	2763
54	197983	3666	2750
55	200721	3649	2737
56	203446	3632	2725
57	206159	3616	2713
58	208861	3601	2701
59	211550	3585	2689
60	214228	3570	2678
61	216895	3555	2666
62	219551	3541	2655
63	222196	3526	2645
64	224830	3512	2634
65	227454	3499	2624
66	230069	3485	2614
67	232673	3472	2604
68	235267	3459	2594
69	237852	3447	2584
70	240427	3434	2575
71	242993	3422	2566
72	245550	3410	2557
73	248098	3398	2548
74	250638	3387	2539
75	253169	3375	2530
76	255691	3364	2522
77	258205	3353	2513
78	260711	3342	2505
79	263208	3331	2497
80	265698	3321	2489
81	268180	3310	2481
82	270654	3300	2474
83	273121	3290	2466
84	275580	3280	2459
85	278032	3270	2451
86	280476	3261	2444
87	282914	3251	2437
88	285344	3242	2430
89	287768	3233	2423
90	290184	3224	2416
91	292594	3215	2409
92	294998	3206	2403
93	297394	3197	2396
94	299784	3189	2390
95	302168	3180	2383
96	304546	3172	2377
97	306917	3164	2371
98	309282	3155	2365
99	311641	3147	2359
100	313994	3139	2353

84%   51 - 100

Appendix II

85 %   LEARNING-CURVE TABLES

UNIT	T C	T A	T U
1	10000	10000	10000
2	17000	8500	7000
3	23187	7729	6187
4	28900	7225	5712
5	34283	6856	5383
6	39418	6569	5135
7	44355	6336	4937
8	49129	6141	4774
9	53765	5973	4635
10	58282	5828	4516
11	62693	5699	4411
12	67011	5584	4318
13	71246	5480	4234
14	75405	5386	4158
15	79494	5299	4089
16	83520	5220	4026
17	87488	5146	3967
18	91401	5077	3913
19	95264	5013	3862
20	99079	4953	3815
21	102850	4897	3770
22	106578	4844	3728
23	110268	4794	3689
24	113919	4746	3651
25	117536	4701	3616
26	121118	4658	3582
27	124669	4617	3550
28	128188	4578	3519
29	131678	4540	3490
30	135141	4504	3462
31	138576	4470	3435
32	141985	4437	3409
33	145370	4405	3384
34	148730	4374	3360
35	152067	4344	3337
36	155383	4316	3315
37	158676	4288	3293
38	161949	4261	3272
39	165201	4235	3252
40	168435	4210	3233
41	171649	4186	3214
42	174845	4162	3195
43	178023	4140	3178
44	181184	4117	3160
45	184328	4096	3144
46	187455	4075	3127
47	190567	4054	3111
48	193663	4034	3096
49	196744	4015	3081
50	199811	3996	3066

85%   1 - 50

Appendix II

LEARNING-CURVE TABLES 85%

UNIT	T C	T A	T U
51	202863	3977	3052
52	205901	3959	3038
53	208926	3942	3024
54	211937	3924	3011
55	214935	3907	2998
56	217920	3891	2985
57	220893	3875	2972
58	223854	3859	2960
59	226802	3844	2948
60	229739	3828	2936
61	232665	3814	2925
62	235579	3799	2914
63	238483	3785	2903
64	241375	3771	2892
65	244257	3757	2881
66	247129	3744	2871
67	249990	3731	2861
68	252841	3718	2851
69	255683	3705	2841
70	258515	3693	2831
71	261337	3680	2822
72	264151	3668	2813
73	266955	3656	2804
74	269750	3645	2795
75	272536	3633	2786
76	275313	3622	2777
77	278082	3611	2768
78	280843	3600	2760
79	283595	3589	2752
80	286339	3579	2744
81	289075	3568	2736
82	291803	3558	2728
83	294524	3548	2720
84	297236	3538	2712
85	299941	3528	2705
86	302639	3519	2697
87	305329	3509	2690
88	308012	3500	2683
89	310688	3490	2675
90	313357	3481	2668
91	316019	3472	2661
92	318674	3463	2655
93	321323	3455	2648
94	323964	3446	2641
95	326599	3437	2635
96	329228	3429	2628
97	331850	3421	2622
98	334466	3412	2615
99	337076	3404	2609
100	339679	3396	2603

86%   LEARNING-CURVE TABLES

UNIT	T C	T A	T U
1	10000	10000	10000
2	17200	8600	7200
3	23621	7873	6421
4	29584	7396	5962
5	35227	7045	5643
6	40628	6771	5401
7	45836	6548	5207
8	50884	6360	5047
9	55796	6199	4912
10	60591	6059	4794
11	65282	5934	4691
12	69881	5823	4599
13	74397	5722	4516
14	78838	5631	4441
15	83211	5547	4372
16	87521	5470	4309
17	91772	5398	4251
18	95970	5331	4197
19	100117	5269	4146
20	104216	5210	4099
21	108271	5155	4055
22	112285	5103	4013
23	116259	5054	3973
24	120195	5008	3936
25	124096	4963	3900
26	127963	4921	3867
27	131798	4881	3834
28	135602	4842	3804
29	139377	4806	3774
30	143124	4770	3746
31	146843	4736	3719
32	150536	4704	3693
33	154205	4672	3668
34	157849	4642	3644
35	161470	4613	3620
36	165068	4585	3598
37	168645	4557	3576
38	172201	4531	3555
39	175736	4506	3535
40	179252	4481	3515
41	182749	4457	3496
42	186227	4433	3478
43	189688	4411	3460
44	193130	4389	3442
45	196556	4367	3425
46	199966	4347	3409
47	203359	4326	3393
48	206736	4307	3377
49	210099	4287	3362
50	213446	4268	3347

Appendix II

LEARNING-CURVE TABLES   86%

UNIT	T C	T A	T U
51	216779	4250	3332
52	220097	4232	3318
53	223402	4215	3304
54	226693	4198	3291
55	229971	4181	3278
56	233237	4164	3265
57	236489	4148	3252
58	239729	4133	3240
59	242957	4117	3227
60	246173	4102	3216
61	249377	4088	3204
62	252570	4073	3192
63	255752	4059	3181
64	258923	4045	3170
65	262083	4032	3160
66	265232	4018	3149
67	268371	4005	3139
68	271500	3992	3128
69	274619	3979	3118
70	277728	3967	3109
71	280828	3955	3099
72	283918	3943	3089
73	286998	3931	3080
74	290070	3919	3071
75	293132	3908	3062
76	296186	3897	3053
77	299231	3886	3044
78	302267	3875	3036
79	305295	3864	3027
80	308314	3853	3019
81	311325	3843	3011
82	314329	3833	3003
83	317324	3823	2995
84	320311	3813	2987
85	323291	3803	2979
86	326263	3793	2972
87	329228	3784	2964
88	332185	3774	2957
89	335135	3765	2949
90	338077	3756	2942
91	341013	3747	2935
92	343941	3738	2928
93	346863	3729	2921
94	349778	3721	2914
95	352686	3712	2908
96	355587	3704	2901
97	358482	3695	2894
98	361370	3687	2888
99	364252	3679	2881
100	367128	3671	2875

86%   51 - 100
                                        Appendix II

87%   LEARNING-CURVE TABLES

UNIT	T C	T A	T U
1	10000	10000	10000
2	17400	8700	7400
3	24058	8019	6658
4	30276	7569	6217
5	36185	7237	5909
6	41861	6976	5675
7	47348	6764	5487
8	52680	6585	5331
9	57879	6431	5199
10	62963	6296	5084
11	67945	6176	4982
12	72838	6069	4892
13	77649	5973	4811
14	82386	5884	4737
15	87056	5803	4669
16	91663	5728	4607
17	96213	5659	4549
18	100709	5594	4496
19	105156	5534	4446
20	109556	5477	4399
21	113911	5424	4355
22	118226	5373	4314
23	122501	5326	4274
24	126738	5280	4237
25	130941	5237	4202
26	135110	5196	4168
27	139246	5157	4136
28	143352	5119	4106
29	147429	5083	4076
30	151477	5049	4048
31	155499	5016	4021
32	159494	4984	3995
33	163465	4953	3970
34	167411	4923	3946
35	171334	4895	3923
36	175235	4867	3900
37	179114	4840	3878
38	182972	4815	3857
39	186809	4789	3837
40	190627	4765	3817
41	194426	4742	3798
42	198206	4719	3780
43	201968	4696	3762
44	205713	4675	3744
45	209440	4654	3727
46	213151	4633	3710
47	216846	4613	3694
48	220525	4594	3678
49	224189	4575	3663
50	227837	4556	3648

Appendix II

LEARNING-CURVE TABLES   87%

UNIT	T C	T A	T U
51	231471	4538	3633
52	235091	4520	3619
53	238697	4503	3605
54	242289	4486	3592
55	245867	4470	3578
56	249433	4454	3565
57	252986	4438	3552
58	256527	4422	3540
59	260055	4407	3528
60	263571	4392	3516
61	267075	4378	3504
62	270568	4364	3492
63	274050	4350	3481
64	277520	4336	3470
65	280980	4322	3459
66	284429	4309	3448
67	287868	4296	3438
68	291296	4283	3428
69	294714	4271	3418
70	298122	4258	3408
71	301520	4246	3398
72	304909	4234	3388
73	308288	4223	3379
74	311659	4211	3370
75	315019	4200	3360
76	318371	4189	3351
77	321714	4178	3343
78	325049	4167	3334
79	328374	4156	3325
80	331692	4146	3317
81	335001	4135	3309
82	338302	4125	3300
83	341594	4115	3292
84	344879	4105	3284
85	348156	4095	3276
86	351425	4086	3269
87	354687	4076	3261
88	357941	4067	3254
89	361187	4058	3246
90	364427	4049	3239
91	367659	4040	3232
92	370884	4031	3224
93	374102	4022	3217
94	377313	4013	3210
95	380517	4005	3204
96	383714	3997	3197
97	386905	3988	3190
98	390089	3980	3184
99	393266	3972	3177
100	396437	3964	3171

87%   51 – 100

Appendix II

88% LEARNING-CURVE TABLES

UNIT	T C	T A	T U
1	10000	10000	10000
2	17600	8800	7600
3	24497	8165	6897
4	30976	7744	6478
5	37158	7431	6182
6	43116	7186	5957
7	48892	6984	5776
8	54517	6814	5625
9	60014	6668	5496
10	65399	6539	5385
11	70686	6426	5286
12	75884	6323	5198
13	81003	6231	5119
14	86050	6146	5046
15	91031	6068	4980
16	95951	5996	4919
17	100814	5930	4863
18	105625	5868	4810
19	110387	5809	4761
20	115103	5755	4715
21	119775	5703	4672
22	124407	5654	4631
23	129000	5608	4593
24	133556	5564	4556
25	138078	5523	4521
26	142566	5483	4488
27	147023	5445	4456
28	151449	5408	4426
29	155846	5374	4397
30	160215	5340	4369
31	164557	5308	4342
32	168874	5277	4316
33	173166	5247	4291
34	177433	5218	4267
35	181678	5190	4244
36	185901	5163	4222
37	190102	5137	4200
38	194282	5112	4180
39	198441	5088	4159
40	202582	5064	4140
41	206703	5041	4121
42	210805	5019	4102
43	214890	4997	4084
44	218957	4976	4067
45	223007	4955	4050
46	227041	4935	4033
47	231058	4916	4017
48	235060	4897	4001
49	239046	4878	3986
50	243017	4860	3971

Appendix II

LEARNING–CURVE TABLES   88%

UNIT	T C	T A	T U
51	246974	4842	3956
52	250917	4825	3942
53	254845	4808	3928
54	258760	4791	3914
55	262661	4775	3901
56	266550	4759	3888
57	270426	4744	3875
58	274289	4729	3863
59	278140	4714	3850
60	281978	4699	3838
61	285805	4685	3827
62	289621	4671	3815
63	293425	4657	3804
64	297218	4644	3793
65	301000	4630	3782
66	304772	4617	3771
67	308533	4604	3760
68	312283	4592	3750
69	316024	4580	3740
70	319754	4567	3730
71	323475	4555	3720
72	327186	4544	3710
73	330887	4532	3701
74	334579	4521	3692
75	338262	4510	3682
76	341936	4499	3673
77	345601	4488	3664
78	349257	4477	3656
79	352905	4467	3647
80	356544	4456	3639
81	360175	4446	3630
82	363797	4436	3622
83	367411	4426	3614
84	371018	4416	3606
85	374616	4407	3598
86	378207	4397	3590
87	381789	4388	3582
88	385365	4379	3575
89	388933	4370	3567
90	392493	4361	3560
91	396046	4352	3553
92	399592	4343	3545
93	403131	4334	3538
94	406663	4326	3531
95	410188	4317	3524
96	413706	4309	3518
97	417217	4301	3511
98	420722	4293	3504
99	424220	4285	3498
100	427711	4277	3491

89%   LEARNING-CURVE TABLES

UNIT	T C	T A	T U
1	10000	10000	10000
2	17800	8900	7800
3	24940	8313	7140
4	31684	7921	6743
5	38146	7629	6462
6	44394	7399	6247
7	50468	7209	6073
8	56397	7049	5929
9	62203	6911	5805
10	67901	6790	5697
11	73503	6682	5602
12	79021	6585	5517
13	84462	6497	5440
14	89833	6416	5370
15	95140	6342	5306
16	100387	6274	5247
17	105580	6210	5192
18	110721	6151	5141
19	115815	6095	5093
20	120864	6043	5048
21	125870	5993	5006
22	130837	5947	4966
23	135765	5902	4928
24	140658	5860	4892
25	145517	5820	4858
26	150343	5782	4826
27	155138	5745	4794
28	159903	5710	4765
29	164640	5677	4736
30	169349	5644	4709
31	174032	5613	4682
32	178689	5584	4657
33	183323	5555	4633
34	187932	5527	4609
35	192519	5500	4586
36	197084	5474	4564
37	201628	5449	4543
38	206151	5425	4523
39	210654	5401	4503
40	215138	5378	4483
41	219602	5356	4464
42	224049	5334	4446
43	228478	5313	4428
44	232889	5292	4411
45	237284	5272	4394
46	241663	5253	4378
47	246025	5234	4362
48	250372	5216	4346
49	254703	5198	4331
50	259020	5180	4316

89%   1 - 50

Appendix II

LEARNING-CURVE TABLES   89%

UNIT	T C	T A	T U
5 1	2 6 3 3 2 2	5 1 6 3	4 3 0 2
5 2	2 6 7 6 1 1	5 1 4 6	4 2 8 8
5 3	2 7 1 8 8 5	5 1 2 9	4 2 7 4
5 4	2 7 6 1 4 6	5 1 1 3	4 2 6 0
5 5	2 8 0 3 9 3	5 0 9 8	4 2 4 7
5 6	2 8 4 6 2 8	5 0 8 2	4 2 3 4
5 7	2 8 8 8 4 9	5 0 6 7	4 2 2 1
5 8	2 9 3 0 5 9	5 0 5 2	4 2 0 9
5 9	2 9 7 2 5 6	5 0 3 8	4 1 9 7
6 0	3 0 1 4 4 1	5 0 2 4	4 1 8 5
6 1	3 0 5 6 1 5	5 0 1 0	4 1 7 3
6 2	3 0 9 7 7 7	4 9 9 6	4 1 6 2
6 3	3 1 3 9 2 8	4 9 8 2	4 1 5 0
6 4	3 1 8 0 6 8	4 9 6 9	4 1 3 9
6 5	3 2 2 1 9 6	4 9 5 6	4 1 2 8
6 6	3 2 6 3 1 5	4 9 4 4	4 1 1 8
6 7	3 3 0 4 2 2	4 9 3 1	4 1 0 7
6 8	3 3 4 5 2 0	4 9 1 9	4 0 9 7
6 9	3 3 8 6 0 7	4 9 0 7	4 0 8 7
7 0	3 4 2 6 8 4	4 8 9 5	4 0 7 7
7 1	3 4 6 7 5 2	4 8 8 3	4 0 6 7
7 2	3 5 0 8 1 0	4 8 7 2	4 0 5 7
7 3	3 5 4 8 5 9	4 8 6 1	4 0 4 8
7 4	3 5 8 8 9 8	4 8 4 9	4 0 3 9
7 5	3 6 2 9 2 8	4 8 3 9	4 0 3 0
7 6	3 6 6 9 4 9	4 8 2 8	4 0 2 1
7 7	3 7 0 9 6 1	4 8 1 7	4 0 1 2
7 8	3 7 4 9 6 4	4 8 0 7	4 0 0 3
7 9	3 7 8 9 5 9	4 7 9 6	3 9 9 4
8 0	3 8 2 9 4 5	4 7 8 6	3 9 8 6
8 1	3 8 6 9 2 3	4 7 7 6	3 9 7 7
8 2	3 9 0 8 9 3	4 7 6 6	3 9 6 9
8 3	3 9 4 8 5 4	4 7 5 7	3 9 6 1
8 4	3 9 8 8 0 8	4 7 4 7	3 9 5 3
8 5	4 0 2 7 5 3	4 7 3 8	3 9 4 5
8 6	4 0 6 6 9 1	4 7 2 8	3 9 3 7
8 7	4 1 0 6 2 1	4 7 1 9	3 9 3 0
8 8	4 1 4 5 4 4	4 7 1 0	3 9 2 2
8 9	4 1 8 4 5 9	4 7 0 1	3 9 1 5
9 0	4 2 2 3 6 6	4 6 9 2	3 9 0 7
9 1	4 2 6 2 6 7	4 6 8 4	3 9 0 0
9 2	4 3 0 1 6 0	4 6 7 5	3 8 9 3
9 3	4 3 4 0 4 6	4 6 6 7	3 8 8 6
9 4	4 3 7 9 2 5	4 6 5 8	3 8 7 9
9 5	4 4 1 7 9 7	4 6 5 0	3 8 7 2
9 6	4 4 5 6 6 2	4 6 4 2	3 8 6 5
9 7	4 4 9 5 2 1	4 6 3 4	3 8 5 8
9 8	4 5 3 3 7 2	4 6 2 6	3 8 5 1
9 9	4 5 7 2 1 8	4 6 1 8	3 8 4 5
1 0 0	4 6 1 0 5 6	4 6 1 0	3 8 3 8

89%

51 - 100
Appendix II

90%   LEARNING-CURVE TABLES

UNIT	T C	T A	T U
1	10000	10000	10000
2	18000	9000	8000
3	25386	8462	7386
4	32400	8100	7013
5	39149	7829	6749
6	45695	7615	6545
7	52076	7439	6381
8	58319	7289	6243
9	64445	7160	6125
10	70468	7046	6023
11	76400	6945	5931
12	82251	6854	5850
13	88027	6771	5776
14	93737	6695	5709
15	99385	6625	5647
16	104975	6560	5590
17	11.0513	6500	5537
18	116002	6444	5488
19	121444	6391	5442
20	126843	6342	5399
21	132201	6295	5358
22	137521	6250	5319
23	142804	6208	5282
24	148052	6168	5248
25	153267	6130	5214
26	158450	6094	5183
27	163603	6059	5153
28	168727	6025	5124
29	173823	5993	5096
30	178893	5963	5069
31	183937	5933	5044
32	188956	5904	5019
33	193952	5877	4995
34	198924	5850	4972
35	203875	5825	4950
36	208804	5800	4928
37	213712	5776	4908
38	218600	5752	4888
39	223469	5729	4868
40	228318	5707	4849
41	233150	5686	4831
42	237963	5665	4813
43	242759	5645	4795
44	247538	5625	4779
45	252301	5606	4762
46	257047	5587	4746
47	261778	5569	4730
48	266493	5551	4715
49	271194	5534	4700
50	275880	5517	4686

90%   1 - 50

Appendix II

LEARNING-CURVE TABLES   90%

UNIT	T C	T A	T U
5 1	2 8 0 5 5 2	5 5 0 1	4 6 7 1
5 2	2 8 5 2 1 0	5 4 8 4	4 6 5 7
5 3	2 8 9 8 5 4	5 4 6 8	4 6 4 4
5 4	2 9 4 4 8 5	5 4 5 3	4 6 3 1
5 5	2 9 9 1 0 3	5 4 3 8	4 6 1 8
5 6	3 0 3 7 0 9	5 4 2 3	4 6 0 5
5 7	3 0 8 3 0 2	5 4 0 8	4 5 9 2
5 8	3 1 2 8 8 2	5 3 9 4	4 5 8 0
5 9	3 1 7 4 5 1	5 3 8 0	4 5 6 8
6 0	3 2 2 0 0 8	5 3 6 6	4 5 5 6
6 1	3 2 6 5 5 3	5 3 5 3	4 5 4 5
6 2	3 3 1 0 8 7	5 3 4 0	4 5 3 3
6 3	3 3 5 6 1 0	5 3 2 7	4 5 2 2
6 4	3 4 0 1 2 2	5 3 1 4	4 5 1 1
6 5	3 4 4 6 2 3	5 3 0 1	4 5 0 1
6 6	3 4 9 1 1 4	5 2 8 9	4 4 9 0
6 7	3 5 3 5 9 4	5 2 7 7	4 4 8 0
6 8	3 5 8 0 6 5	5 2 6 5	4 4 7 0
6 9	3 6 2 5 2 5	5 2 5 3	4 4 6 0
7 0	3 6 6 9 7 5	5 2 4 2	4 4 5 0
7 1	3 7 1 4 1 6	5 2 3 1	4 4 4 0
7 2	3 7 5 8 4 7	5 2 2 0	4 4 3 1
7 3	3 8 0 2 6 9	5 2 0 9	4 4 2 1
7 4	3 8 4 6 8 2	5 1 9 8	4 4 1 2
7 5	3 8 9 0 8 6	5 1 8 7	4 4 0 3
7 6	3 9 3 4 8 1	5 1 7 7	4 3 9 4
7 7	3 9 7 8 6 7	5 1 6 7	4 3 8 6
7 8	4 0 2 2 4 4	5 1 5 6	4 3 7 7
7 9	4 0 6 6 1 3	5 1 4 7	4 3 6 8
8 0	4 1 0 9 7 4	5 1 3 7	4 3 6 0
8 1	4 1 5 3 2 6	5 1 2 7	4 3 5 2
8 2	4 1 9 6 7 0	5 1 1 7	4 3 4 4
8 3	4 2 4 0 0 6	5 1 0 8	4 3 3 5
8 4	4 2 8 3 3 4	5 0 9 9	4 3 2 8
8 5	4 3 2 6 5 4	5 0 9 0	4 3 2 0
8 6	4 3 6 9 6 7	5 0 8 1	4 3 1 2
8 7	4 4 1 2 7 1	5 0 7 2	4 3 0 4
8 8	4 4 5 5 6 9	5 0 6 3	4 2 9 7
8 9	4 4 9 8 5 9	5 0 5 4	4 2 8 9
9 0	4 5 4 1 4 1	5 0 4 6	4 2 8 2
9 1	4 5 8 4 1 7	5 0 3 7	4 2 7 5
9 2	4 6 2 6 8 5	5 0 2 9	4 2 6 8
9 3	4 6 6 9 4 6	5 0 2 0	4 2 6 1
9 4	4 7 1 2 0 1	5 0 1 2	4 2 5 4
9 5	4 7 5 4 4 8	5 0 0 4	4 2 4 7
9 6	4 7 9 6 8 9	4 9 9 6	4 2 4 0
9 7	4 8 3 9 2 2	4 9 8 8	4 2 3 3
9 8	4 8 8 1 5 0	4 9 8 1	4 2 2 7
9 9	4 9 2 3 7 0	4 9 7 3	4 2 2 0
1 0 0	4 9 6 5 8 5	4 9 6 5	4 2 1 4

90%   51 - 100
                                   Appendix II

91%   LEARNING-CURVE TABLES

UNIT	T C	T A	T U
1	10000	10000	10000
2	18200	9100	8200
3	25834	8611	7634
4	33124	8281	7289
5	40166	8033	7042
6	47019	7836	6852
7	53717	7673	6697
8	60285	7535	6568
9	66743	7415	6457
10	73103	7310	6360
11	79377	7216	6274
12	85574	7131	6197
13	91701	7053	6127
14	97765	6983	6063
15	103769	6917	6004
16	109719	6857	5950
17	115619	6801	5899
18	121472	6748	5852
19	127281	6699	5808
20	133048	6652	5767
21	138776	6608	5728
22	144467	6566	5691
23	150123	6527	5655
24	155746	6489	5622
25	161337	6453	5590
26	166897	6419	5560
27	172428	6386	5531
28	177932	6354	5503
29	183409	6324	5476
30	188860	6295	5451
31	194287	6267	5426
32	199690	6240	5402
33	205070	6214	5379
34	210427	6189	5357
35	215764	6164	5336
36	221080	6141	5315
37	226375	6118	5295
38	231651	6096	5276
39	236909	6074	5257
40	242148	6053	5239
41	247369	6033	5221
42	252573	6013	5203
43	257760	5994	5187
44	262931	5975	5170
45	268085	5957	5154
46	273225	5939	5139
47	278349	5922	5123
48	283458	5905	5109
49	288552	5888	5094
50	293633	5872	5080

91%   1 - 50

Appendix II

LEARNING-CURVE TABLES 91%

UNIT	T C	T A	T U
51	298700	5856	5066
52	303753	5841	5053
53	308793	5826	5040
54	313820	5811	5027
55	318835	5797	5014
56	323837	5782	5002
57	328827	5768	4989
58	333805	5755	4978
59	338771	5741	4966
60	343726	5728	4954
61	348670	5715	4943
62	353603	5703	4932
63	358525	5690	4921
64	363436	5678	4911
65	368337	5666	4900
66	373227	5654	4890
67	378108	5643	4880
68	382979	5632	4870
69	387839	5620	4860
70	392691	5609	4851
71	397533	5599	4841
72	402365	5588	4832
73	407189	5577	4823
74	412003	5567	4814
75	416809	5557	4805
76	421606	5547	4796
77	426394	5537	4788
78	431174	5527	4779
79	435946	5518	4771
80	440709	5508	4763
81	445465	5499	4755
82	450212	5490	4747
83	454951	5481	4739
84	459683	5472	4731
85	464407	5463	4724
86	469124	5454	4716
87	473833	5446	4709
88	478534	5437	4701
89	483229	5429	4694
90	487916	5421	4687
91	492596	5413	4680
92	497269	5405	4673
93	501935	5397	4666
94	506595	5389	4659
95	511247	5381	4652
96	515893	5373	4646
97	520533	5366	4639
98	525166	5358	4632
99	529792	5351	4626
100	534412	5344	4620

91%   51 - 100

92%   LEARNING-CURVE TABLES

UNIT	T C	T A	T U
1	10000	10000	10000
2	18400	9200	8400
3	26286	8762	7886
4	33856	8464	7569
5	41199	8239	7343
6	48366	8061	7167
7	55390	7912	7024
8	62295	7786	6904
9	69095	7677	6800
10	75806	7580	6710
11	82436	7494	6630
12	88994	7416	6557
13	95486	7345	6492
14	101919	7279	6432
15	108296	7219	6377
16	114622	7163	6326
17	120901	7111	6279
18	127136	7063	6234
19	133329	7017	6193
20	139483	6974	6154
21	145600	6933	6117
22	151683	6894	6082
23	157731	6857	6048
24	163749	6822	6017
25	169736	6789	5987
26	175695	6757	5958
27	181626	6726	5931
28	187531	6697	5904
29	193410	6669	5879
30	199265	6642	5855
31	205096	6616	5831
32	210905	6590	5809
33	216693	6566	5787
34	222459	6542	5766
35	228205	6520	5745
36	233931	6498	5726
37	239638	6476	5706
38	245326	6455	5688
39	250996	6435	5670
40	256649	6416	5652
41	262285	6397	5636
42	267905	6378	5619
43	273508	6360	5603
44	279096	6343	5587
45	284669	6325	5572
46	290226	6309	5557
47	295769	6292	5543
48	301298	6277	5528
49	306813	6261	5515
50	312315	6246	5501

Appendix II

LEARNING-CURVE TABLES   92%

UNIT	T C	T A	T U
51	317803	6231	5488
52	323279	6216	5475
53	328742	6202	5462
54	334192	6188	5450
55	339630	6175	5438
56	345057	6161	5426
57	350471	6148	5414
58	355875	6135	5403
59	361267	6123	5392
60	366648	6110	5381
61	372018	6098	5370
62	377378	6086	5359
63	382727	6075	5349
64	388067	6063	5339
65	393396	6052	5329
66	398715	6041	5319
67	404025	6030	5309
68	409325	6019	5300
69	414615	6008	5290
70	419897	5998	5281
71	425169	5988	5272
72	430433	5978	5263
73	435688	5968	5254
74	440934	5958	5246
75	446171	5948	5237
76	451400	5939	5229
77	456621	5930	5220
78	461834	5920	5212
79	467039	5911	5204
80	472235	5902	5196
81	477424	5894	5188
82	482606	5885	5181
83	487779	5876	5173
84	492946	5868	5166
85	498104	5860	5158
86	503256	5851	5151
87	508400	5843	5144
88	513537	5835	5137
89	518667	5827	5130
90	523791	5819	5123
91	528907	5812	5116
92	534017	5804	5109
93	539120	5796	5102
94	544216	5789	5096
95	549306	5782	5089
96	554389	5774	5083
97	559466	5767	5077
98	564537	5760	5070
99	569602	5753	5064
100	574660	5746	5058

92%   51 - 100

93%   LEARNING-CURVE TABLES

UNIT	T C	T A	T U
1	10000	10000	10000
2	18600	9300	8600
3	26740	8913	8140
4	34596	8649	7855
5	42246	8449	7650
6	49737	8289	7490
7	57097	8156	7360
8	64348	8043	7250
9	71504	7944	7156
10	78578	7857	7073
11	85577	7779	6999
12	92511	7709	6933
13	99383	7644	6872
14	106201	7585	6817
15	112968	7531	6766
16	119688	7480	6719
17	126364	7433	6675
18	132999	7388	6634
19	139595	7347	6596
20	146155	7307	6560
21	152681	7270	6525
22	159174	7235	6493
23	165637	7201	6462
24	172070	7169	6433
25	178475	7139	6405
26	184854	7109	6378
27	191206	7081	6352
28	197535	7054	6328
29	203839	7028	6304
30	210121	7004	6281
31	216381	6980	6259
32	222620	6956	6238
33	228838	6934	6218
34	235037	6912	6198
35	241216	6891	6179
36	247378	6871	6161
37	253521	6851	6143
38	259647	6832	6125
39	265756	6814	6109
40	271849	6796	6092
41	277926	6778	6076
42	283987	6761	6061
43	290033	6744	6046
44	296065	6728	6031
45	302082	6712	6017
46	308085	6697	6003
47	314074	6682	5989
48	320051	6667	5976
49	326014	6653	5963
50	331964	6639	5950

93%   1 - 50

Appendix II

LEARNING-CURVE TABLES   93%

UNIT	T C	T A	T U
5 1	3 3 7 9 0 2	6 6 2 5	5 9 3 8
5 2	3 4 3 8 2 8	6 6 1 2	5 9 2 5
5 3	3 4 9 7 4 2	6 5 9 8	5 9 1 3
5 4	3 5 5 6 4 4	6 5 8 6	5 9 0 2
5 5	3 6 1 5 3 5	6 5 7 3	5 8 9 0
5 6	3 6 7 4 1 5	6 5 6 0	5 8 7 9
5 7	3 7 3 2 8 3	6 5 4 8	5 8 6 8
5 8	3 7 9 1 4 1	6 5 3 6	5 8 5 7
5 9	3 8 4 9 8 8	6 5 2 5	5 8 4 7
6 0	3 9 0 8 2 5	6 5 1 3	5 8 3 6
6 1	3 9 6 6 5 2	6 5 0 2	5 8 2 6
6 2	4 0 2 4 6 9	6 4 9 1	5 8 1 6
6 3	4 0 8 2 7 6	6 4 8 0	5 8 0 6
6 4	4 1 4 0 7 3	6 4 6 9	5 7 9 7
6 5	4 1 9 8 6 1	6 4 5 9	5 7 8 7
6 6	4 2 5 6 3 9	6 4 4 9	5 7 7 8
6 7	4 3 1 4 0 9	6 4 3 8	5 7 6 9
6 8	4 3 7 1 6 9	6 4 2 8	5 7 6 0
6 9	4 4 2 9 2 0	6 4 1 9	5 7 5 1
7 0	4 4 8 6 6 3	6 4 0 9	5 7 4 2
7 1	4 5 4 3 9 7	6 3 9 9	5 7 3 4
7 2	4 6 0 1 2 3	6 3 9 0	5 7 2 5
7 3	4 6 5 8 4 0	6 3 8 1	5 7 1 7
7 4	4 7 1 5 5 0	6 3 7 2	5 7 0 9
7 5	4 7 7 2 5 1	6 3 6 3	5 7 0 1
7 6	4 8 2 9 4 4	6 3 5 4	5 6 9 3
7 7	4 8 8 6 2 9	6 3 4 5	5 6 8 5
7 8	4 9 4 3 0 7	6 3 3 7	5 6 7 7
7 9	4 9 9 9 7 7	6 3 2 8	5 6 6 9
8 0	5 0 5 6 3 9	6 3 2 0	5 6 6 2
8 1	5 1 1 2 9 4	6 3 1 2	5 6 5 5
8 2	5 1 6 9 4 2	6 3 0 4	5 6 4 7
8 3	5 2 2 5 8 3	6 2 9 6	5 6 4 0
8 4	5 2 8 2 1 6	6 2 8 8	5 6 3 3
8 5	5 3 3 8 4 2	6 2 8 0	5 6 2 6
8 6	5 3 9 4 6 2	6 2 7 2	5 6 1 9
8 7	5 4 5 0 7 5	6 2 6 5	5 6 1 2
8 8	5 5 0 6 8 0	6 2 5 7	5 6 0 5
8 9	5 5 6 2 8 0	6 2 5 0	5 5 9 9
9 0	5 6 1 8 7 2	6 2 4 3	5 5 9 2
9 1	5 6 7 4 5 9	6 2 3 5	5 5 8 6
9 2	5 7 3 0 3 8	6 2 2 8	5 5 7 9
9 3	5 7 8 6 1 2	6 2 2 1	5 5 7 3
9 4	5 8 4 1 7 9	6 2 1 4	5 5 6 7
9 5	5 8 9 7 4 0	6 2 0 7	5 5 6 0
9 6	5 9 5 2 9 5	6 2 0 0	5 5 5 4
9 7	6 0 0 8 4 3	6 1 9 4	5 5 4 8
9 8	6 0 6 3 8 6	6 1 8 7	5 5 4 2
9 9	6 1 1 9 2 3	6 1 8 1	5 5 3 6
1 0 0	6 1 7 4 5 4	6 1 7 4	5 5 3 0

94%   LEARNING-CURVE TABLES

UNIT	T C	T A	T U
1	10000	10000	10000
2	18800	9400	8800
3	27197	9065	8397
4	35344	8836	8146
5	43308	8661	7964
6	51131	8521	7822
7	58838	8405	7706
8	66446	8305	7608
9	73970	8218	7524
10	81420	8142	7449
11	88803	8073	7383
12	96127	8010	7323
13	103396	7953	7269
14	110615	7901	7219
15	117789	7852	7173
16	124919	7807	7130
17	132011	7765	7091
18	139065	7725	7053
19	146084	7688	7019
20	153070	7653	6986
21	160025	7620	6955
22	166950	7588	6925
23	173848	7558	6897
24	180718	7529	6870
25	187564	7502	6845
26	194384	7476	6820
27	201182	7451	6797
28	207957	7427	6774
29	214710	7403	6753
30	221443	7381	6732
31	228156	7359	6712
32	234849	7339	6693
33	241524	7318	6674
34	248180	7299	6656
35	254819	7280	6639
36	261442	7262	6622
37	268048	7244	6605
38	274638	7227	6589
39	281212	7210	6574
40	287772	7194	6559
41	294316	7178	6544
42	300847	7163	6530
43	307364	7148	6516
44	313867	7133	6503
45	320357	7119	6490
46	326834	7105	6477
47	333299	7091	6464
48	339751	7078	6452
49	346192	7065	6440
50	352620	7052	6428

94%   1 - 50

Appendix II

LEARNING-CURVE TABLES 94%

UNIT	T C	T A	T U
5 1	3 5 9 0 3 7	7 0 3 9	6 4 1 7
5 2	3 6 5 4 4 3	7 0 2 7	6 4 0 5
5 3	3 7 1 8 3 8	7 0 1 5	6 3 9 4
5 4	3 7 8 2 2 2	7 0 0 4	6 3 8 4
5 5	3 8 4 5 9 6	6 9 9 2	6 3 7 3
5 6	3 9 0 9 5 9	6 9 8 1	6 3 6 3
5 7	3 9 7 3 1 3	6 9 7 0	6 3 5 3
5 8	4 0 3 6 5 6	6 9 5 9	6 3 4 3
5 9	4 0 9 9 8 9	6 9 4 8	6 3 3 3
6 0	4 1 6 3 1 3	6 9 3 8	6 3 2 3
6 1	4 2 2 6 2 8	6 9 2 8	6 3 1 4
6 2	4 2 8 9 3 3	6 9 1 8	6 3 0 5
6 3	4 3 5 2 2 9	6 9 0 8	6 2 9 6
6 4	4 4 1 5 1 6	6 8 9 8	6 2 8 7
6 5	4 4 7 7 9 5	6 8 8 9	6 2 7 8
6 6	4 5 4 0 6 5	6 8 7 9	6 2 6 9
6 7	4 6 0 3 2 6	6 8 7 0	6 2 6 1
6 8	4 6 6 5 7 9	6 8 6 1	6 2 5 3
6 9	4 7 2 8 2 4	6 8 5 2	6 2 4 4
7 0	4 7 9 0 6 1	6 8 4 3	6 2 3 6
7 1	4 8 5 2 9 0	6 8 3 5	6 2 2 8
7 2	4 9 1 5 1 1	6 8 2 6	6 2 2 1
7 3	4 9 7 7 2 4	6 8 1 8	6 2 1 3
7 4	5 0 3 9 3 0	6 8 0 9	6 2 0 5
7 5	5 1 0 1 2 8	6 8 0 1	6 1 9 8
7 6	5 1 6 3 1 9	6 7 9 3	6 1 9 0
7 7	5 2 2 5 0 3	6 7 8 5	6 1 8 3
7 8	5 2 8 6 7 9	6 7 7 7	6 1 7 6
7 9	5 3 4 8 4 9	6 7 7 0	6 1 6 9
8 0	5 4 1 0 1 1	6 7 6 2	6 1 6 2
8 1	5 4 7 1 6 7	6 7 5 5	6 1 5 5
8 2	5 5 3 3 1 5	6 7 4 7	6 1 4 8
8 3	5 5 9 4 5 7	6 7 4 0	6 1 4 2
8 4	5 6 5 5 9 3	6 7 3 3	6 1 3 5
8 5	5 7 1 7 2 2	6 7 2 6	6 1 2 8
8 6	5 7 7 8 4 4	6 7 1 9	6 1 2 2
8 7	5 8 3 9 6 0	6 7 1 2	6 1 1 6
8 8	5 9 0 0 7 0	6 7 0 5	6 1 0 9
8 9	5 9 6 1 7 4	6 6 9 8	6 1 0 3
9 0	6 0 2 2 7 2	6 6 9 1	6 0 9 7
9 1	6 0 8 3 6 3	6 6 8 5	6 0 9 1
9 2	6 1 4 4 4 9	6 6 7 8	6 0 8 5
9 3	6 2 0 5 2 8	6 6 7 2	6 0 7 9
9 4	6 2 6 6 0 2	6 6 6 5	6 0 7 3
9 5	6 3 2 6 7 0	6 6 5 9	6 0 6 8
9 6	6 3 8 7 3 3	6 6 5 3	6 0 6 2
9 7	6 4 4 7 8 9	6 6 4 7	6 0 5 6
9 8	6 5 0 8 4 0	6 6 4 1	6 0 5 1
9 9	6 5 6 8 8 6	6 6 3 5	6 0 4 5
1 0 0	6 6 2 9 2 6	6 6 2 9	6 0 4 0

LEARNING-CURVE TABLES

UNIT	T C	T A	T U
1	10000	10000	10000
2	19000	9500	9000
3	27657	9219	8657
4	36100	9025	8442
5	44385	8877	8285
6	52549	8758	8163
7	60612	8658	8062
8	68589	8573	7977
9	76494	8499	7904
10	84333	8433	7839
11	92114	8374	7781
12	99843	8320	7729
13	107525	8271	7681
14	115163	8225	7637
15	122760	8184	7597
16	130321	8145	7560
17	137846	8108	7525
18	145338	8074	7492
19	152800	8042	7461
20	160233	8011	7432
21	167638	7982	7405
22	175017	7955	7379
23	182372	7929	7354
24	189703	7904	7330
25	197011	7880	7308
26	204298	7857	7286
27	211564	7835	7265
28	218810	7814	7246
29	226037	7794	7226
30	233245	7774	7208
31	240436	7756	7190
32	247609	7737	7173
33	254766	7720	7157
34	261907	7703	7140
35	269033	7686	7125
36	276143	7670	7110
37	283239	7655	7095
38	290321	7640	7081
39	297388	7625	7067
40	304443	7611	7054
41	311484	7597	7041
42	318513	7583	7028
43	325529	7570	7016
44	332534	7557	7004
45	339526	7545	6992
46	346507	7532	6981
47	353477	7520	6969
48	360436	7509	6958
49	367384	7497	6948
50	374321	7486	6937

LEARNING-CURVE TABLES   95%

UNIT	T C	T A	T U
51	381249	7475	6927
52	388166	7464	6917
53	395074	7454	6907
54	401971	7443	6897
55	408860	7433	6888
56	415739	7423	6879
57	422609	7414	6870
58	429470	7404	6861
59	436322	7395	6852
60	443166	7386	6843
61	450001	7377	6835
62	456828	7368	6827
63	463647	7359	6818
64	470458	7350	6810
65	477261	7342	6803
66	484057	7334	6795
67	490844	7326	6787
68	497624	7318	6780
69	504397	7310	6772
70	511163	7302	6765
71	517921	7294	6758
72	524673	7287	6751
73	531417	7279	6744
74	538155	7272	6737
75	544885	7265	6730
76	551610	7258	6724
77	558327	7251	6717
78	565038	7244	6711
79	571743	7237	6704
80	578442	7230	6698
81	585134	7223	6692
82	591821	7217	6686
83	598501	7210	6680
84	605175	7204	6674
85	611844	7198	6668
86	618506	7191	6662
87	625163	7185	6656
88	631814	7179	6651
89	638460	7173	6645
90	645100	7167	6640
91	651735	7161	6634
92	658364	7156	6629
93	664988	7150	6623
94	671606	7144	6618
95	678220	7139	6613
96	684828	7133	6608
97	691431	7128	6603
98	698030	7122	6598
99	704623	7117	6593
100	711211	7112	6588

95%

51 - 100
Appendix II

96%   LEARNING-CURVE TABLES

UNIT	T C	T A	T U
1	10000	10000	10000
2	19200	9600	9200
3	28120	9373	8920
4	36864	9216	8743
5	45478	9095	8614
6	53991	8998	8512
7	62420	8917	8429
8	70778	8847	8358
9	79075	8786	8296
10	87318	8731	8242
11	95512	8682	8194
12	103663	8638	8150
13	111773	8597	8110
14	119847	8560	8073
15	127887	8525	8039
16	135895	8493	8008
17	143874	8463	7978
18	151825	8434	7951
19	159750	8407	7925
20	167651	8382	7900
21	175528	8358	7877
22	183384	8335	7855
23	191218	8313	7834
24	199033	8293	7814
25	206828	8273	7795
26	214605	8254	7776
27	222364	8235	7759
28	230106	8218	7742
29	237832	8201	7726
30	245543	8184	7710
31	253238	8168	7695
32	260919	8153	7680
33	268585	8138	7666
34	276238	8124	7652
35	283878	8110	7639
36	291504	8097	7626
37	299119	8084	7614
38	306721	8071	7602
39	314311	8059	7590
40	321890	8047	7578
41	329458	8035	7567
42	337015	8024	7556
43	344561	8013	7546
44	352098	8002	7536
45	359623	7991	7525
46	367140	7981	7516
47	374646	7971	7506
48	382143	7961	7497
49	389631	7951	7487
50	397110	7942	7478

96%   1 - 50

Appendix II

LEARNING-CURVE TABLES   96%

UNIT	T C	T A	T U
5 1	4 0 4 5 8 0	7 9 3 2	7 4 7 0
5 2	4 1 2 0 4 1	7 9 2 3	7 4 6 1
5 3	4 1 9 4 9 4	7 9 1 4	7 4 5 3
5 4	4 2 6 9 3 9	7 9 0 6	7 4 4 4
5 5	4 3 4 3 7 6	7 8 9 7	7 4 3 6
5 6	4 4 1 8 0 5	7 8 8 9	7 4 2 8
5 7	4 4 9 2 2 5	7 8 8 1	7 4 2 0
5 8	4 5 6 6 3 9	7 8 7 3	7 4 1 3
5 9	4 6 4 0 4 4	7 8 6 5	7 4 0 5
6 0	4 7 1 4 4 2	7 8 5 7	7 3 9 8
6 1	4 7 8 8 3 4	7 8 4 9	7 3 9 1
6 2	4 8 6 2 1 7	7 8 4 2	7 3 8 3
6 3	4 9 3 5 9 4	7 8 3 4	7 3 7 6
6 4	5 0 0 9 6 4	7 8 2 7	7 3 6 9
6 5	5 0 8 3 2 7	7 8 2 0	7 3 6 3
6 6	5 1 5 6 8 4	7 8 1 3	7 3 5 6
6 7	5 2 3 0 3 4	7 8 0 6	7 3 4 9
6 8	5 3 0 3 7 7	7 7 9 9	7 3 4 3
6 9	5 3 7 7 1 5	7 7 9 2	7 3 3 7
7 0	5 4 5 0 4 6	7 7 8 6	7 3 3 0
7 1	5 5 2 3 7 0	7 7 7 9	7 3 2 4
7 2	5 5 9 6 8 9	7 7 7 3	7 3 1 8
7 3	5 6 7 0 0 2	7 7 6 7	7 3 1 2
7 4	5 7 4 3 0 8	7 7 6 0	7 3 0 6
7 5	5 8 1 6 0 9	7 7 5 4	7 3 0 0
7 6	5 8 8 9 0 5	7 7 4 8	7 2 9 5
7 7	5 9 6 1 9 4	7 7 4 2	7 2 8 9
7 8	6 0 3 4 7 8	7 7 3 6	7 2 8 4
7 9	6 1 0 7 5 7	7 7 3 1	7 2 7 8
8 0	6 1 8 0 3 0	7 7 2 5	7 2 7 3
8 1	6 2 5 2 9 7	7 7 1 9	7 2 6 7
8 2	6 3 2 5 6 0	7 7 1 4	7 2 6 2
8 3	6 3 9 8 1 7	7 7 0 8	7 2 5 7
8 4	6 4 7 0 6 9	7 7 0 3	7 2 5 2
8 5	6 5 4 3 1 6	7 6 9 7	7 2 4 7
8 6	6 6 1 5 5 8	7 6 9 2	7 2 4 1
8 7	6 6 8 7 9 5	7 6 8 7	7 2 3 7
8 8	6 7 6 0 2 7	7 6 8 2	7 2 3 2
8 9	6 8 3 2 5 5	7 6 7 7	7 2 2 7
9 0	6 9 0 4 7 7	7 6 7 1	7 2 2 2
9 1	6 9 7 6 9 5	7 6 6 6	7 2 1 7
9 2	7 0 4 9 0 8	7 6 6 2	7 2 1 3
9 3	7 1 2 1 1 7	7 6 5 7	7 2 0 8
9 4	7 1 9 3 2 1	7 6 5 2	7 2 0 3
9 5	7 2 6 5 2 0	7 6 4 7	7 1 9 9
9 6	7 3 3 7 1 5	7 6 4 2	7 1 9 4
9 7	7 4 0 9 0 6	7 6 3 8	7 1 9 0
9 8	7 4 8 0 9 2	7 6 3 3	7 1 8 6
9 9	7 5 5 2 7 4	7 6 2 9	7 1 8 1
1 0 0	7 6 2 4 5 1	7 6 2 4	7 1 7 7

Appendix 3
Method 5—Parametric Estimating Feasibility Model with Typical Charts and Formulas

The two major scientific developments in the control and forecasting of production hours are:

1. Standard hours, which are units of measure that can be applied to direct production labor.
2. Learning curves, which can be used to forecast the actual hours necessary to produce one standard hour of work at any given unit or for any quantity.

Method 5 provides the key methods of using these most important techniques, and unites them so that they can be used at the very beginning of a program.

TYPICAL APPLICATION

To illustrate the use of this system, the following charts and attendant forms are provided for a hypothetical company.

If we assume that this company does mechanical, sheet metal, and electrical (or electronic)* work, we will have the basic charts for organizations doing any one, or a combination, of the three types of work.

The charts are presented as a starting point to make it easier for anyone interested in this method to get started in its use. They do not reflect the average for any industry or firm, and are not recommended for use in their present form. In actual application they must reflect the accepted standard hours for typical operations performed by the respective firms using them. In many cases, however, the forms given can be readily adapted by changing an index point or two, or by the revision of one scale only; others must be completely redone.

A company interested in mechanical work only will find it advisable to make charts for each cost center; the same is true for a firm interested only in sheet metal work, or electrical work. A company using all three types of work will find that additions are required in certain areas. This method provides for such growth, and any additional charts should follow the plan outlined in Chapter 6 so that there is no double coverage.

---

*Throughout this book the general term *electrical* has been used in all discussion and illustrations. It should be understood that its use is in all cases synonymous with *electronic*. No specific differentiation has been made.

Chapter 6 presented only an introduction to this system, providing the key methods of making the standard hour charts. After the standard hours are known, the factors for hours per standard hour are needed to develop the estimated hours for a project. This appendix presents a method of developing these factors, giving due allowance for the major elements that affect them.

Outline

The method, as presented, is outlined as follows:

1. The necessary forms are provided. Two types are needed:
   a. One to summarize the known elements using a method similar to an indentured parts list.
   b. The form on which the estimate is made. It is provided with a line for each type of work and provisions to note the totals from form a.
2. A nomogram is given for estimating the standard hours for each type of work. *These are keyed by number to the lines for the elements on the estimating form.* (Formulas are included also.)
   (NOTE: Numbers 5, 12, 24, and 25 have been left free for growth, and are not used in our example.)
3. The necessary elements to develop the hours per standard hour are next. It is here that a full understanding of learning curves is necessary. (See Chapters 8, 9, and 10.) For each type of work we need hours per standard hour for unit one, and the forecast learning-curve percent.

Our charts consider the elements in item 3, first by total program to illustrate and establish the principle in a simple manner, and then by type of work so that any mix can be estimated. The estimating form has provision for noting these elements in the parenthetical spaces on the form.

A fast method of applying the learning-curve data to develop the hours per standard hour by type of work per project is given in Fig. 3-39.

Theory and Instructions

Method 5 is a method of estimating that follows a definite sequence which is outlined by the estimating form. Each step is presented in a definite manner and may seem to be purely mechanical. This is partially true in the calculation and application of standard hours, but it is not the end result. The development of the estimated hours for a project, from the standard hours, requires that proper consideration be given to many variables, and the best that this method provides is a means of directing one's judgment in considering the same representative group of variables, so that they can be consistently applied to every project.

Procedure

The six major procedure steps are given here, each step being elaborated to help in establishing the use of the method.

1. Secure the best possible bill of material for the product.
    a. There are various sources for a bill of material, such as:
        i. Assembly parts lists.
        ii. Take from prints.
        iii. Maintenance handbooks.
        iv. Advance bills of material. (These may not include all the required material, so care must be used to add the missing items.)
        v. Similar items combined with additions and deletions, if any.
        vi. The engineer's best estimate of the components required. Emphasize the importance of accuracy.

    Each item is critical in determining the estimate. Estimating definite quantities of each item as required will be simple in some areas; in others it will require a great deal of consideration. Often, an error in a certain item, in itself, will not materially affect the estimate, but several errors in the same direction can be critical.

    Special care must be used in determining the values for harnesses. Some of the elements that are useful in this are:

        i. Total number of pins in all plugs.
        ii. Total printed circuit boards with attendant plugs.
        iii. Schematics and wiring diagrams.
        iv. Total lugs on all multiple-terminal components.
        v. Photographs of the item.
        vi. Similar items.

2. Tabulate the bill of material to conform to the listings and values for the mechanical, sheet metal, and electrical work as outlined in the estimating form, and fill in the form with the items as tabulated.
    a. Due to the relatively few categories it is possible to put on a single form, it is necessary for the broadest possible scope to be given to some items so that any project can be estimated. The list tabulated in Fig. A3-1 should be of assistance in determining the proper categories for components that do not appear on the estimating form for want of space.
3. There is a chart (nomogram) plus a formula for each item listed on the estimating form. Using the values for each item, and the respective nomogram, calculate the standard hours for fabrication, assembly, and test, as required. Total these as provided by the form. (This is in preparation for step 4.)
    a. While there is a chart for each item listed on the estimating form, certain limitations must be understood. A system cannot be developed to handle

	Mechanical				Sheet Metal					Electrical										
	1 Castings	2 Gears	3 Machined Parts	4 Bolts; Screws; Etc.	7 Boxes; Racks; Chassis; Etc.	8 Screws; Rivets; Terminals	9 Welding – Heliarc	10 Brazing; Silver Solder	11 Spot Welding	14 Circuit Welding	15 Printed-Circuit Boards	16 Harnesses	17 Plugs – All Types	18 Tubes With Sockets	19 Miniature Tubes	20 Transistors	21 Diodes	22 Multi-Term. Comp.	23 Pig-Tail Components	24
Bearings				X																
Bellows						X														
Black-Boxes					X	X	X	X	X											
Blowers																		X		
Bobbins						X														
Bolts				X																
Brazing								X												
Bushings				X		X														
Cables													X							
Cans					X	X														
Cases					X	X	X	X	X											
Castings	X																			
Capacitors																			X	
Chassis					X															
Chokes																		X		
Choppers																		X		
Circuit Welding										X										
Clamps					X	X	X													
Clips					X	X														
Coils																		X		
Collars			X			X														
Condensers																		X		
Connectors													X							
Contacts						X														
Converters																		X		
Cores						X														
Couplers																		X		
Coupling						X														
Crystals																	X			
Delay Lines																		X		
Dial Controls																		X		
Diodes																	X			
Dust Caps						X														
Eyelets						X														

Fig. A3-1. Supplementary list for determining categories for components not listed on estimating form.

	Mechanical				Sheet Metal					Electrical										
	Castings	Gears	Machined Parts	Bolts; Screws; Etc.	Boxes; Racks; Chassis; Etc.	Screws; Rivets; Terminals	Welding – Heliarc	Brazing; Silver Solder	Spot Welding	Circuit Welding	Printed-Circuit Boards	Harnesses	Plugs – All Types	Tubes With Sockets	Miniature Tubes	Transistors	Diodes	Multi-Term. Comp.	Pig-Tail Components	
	1	2	3	4	7	8	9	10	11	14	15	16	17	18	19	20	21	22	23	24
Fans																		X		
Fasteners						X														
Filters																		X		
Fittings				X		X														
Flanges			X			X														
Forgings	X		X																	
Fuse Holders																			X	
Fuses						X														
Gaskets					X	X														
Gauges																		X		
Gears		X																		
Generators																		X		
Gyros																		X		
Grommets						X														
Handles				X		X														
Harnesses												X								
Heaters																		X		
Holders						X													X	
Impellers																		X		
Inductors																		X		
Inserts						X														
Insulators						X														
Jacks																		X		
Jumpers																			X	
Knobs						X														
Lamps																		X		
Lugs						X														
Machined Parts			X																	
Magnetic Amplifiers																		X		
Meters																		X		
Mechanical, Misc.				X		X														
Motors																		X		
Multi-Term. Comp.																		X		
Nuts (Included with the bolt – Do not count)																				

Fig. A3-1. (Contd.) Supplementary list for determining categories for components not listed on estimating form.

any and every product or project without becoming very massive, and it is not claimed that these charts as given will handle any and all programs. This is presented more as a method, at this time, and these charts should be modified, or new charts added as required. Therefore, if a new product is dissimilar to previous work to such an extent that certain work is unique,

	Mechanical				Sheet Metal								Electrical								
	Castings	Gears	Machined Parts	Bolts; Screws; Etc.	Boxes; Racks; Chassis; Etc.	Screws; Rivets; Terminals	Welding – Heliarc	Brazing; Silver Solder	Spot Welding	Circuit Welding	Printed-Circuit Boards	Harnesses	Plugs – All Types	Tubes With Sockets	Miniature Tubes	Transistors	Diodes	Multi-Term. Comp.	Pig-Tail Components	
	1	2	3	4	7	8	9	10	11	14	15	16	17	18	19	20	21	22	23	24
"O" Rings						X														
Pins						X														
Pig-Tail Components																			X	
Plugs – All Kinds													X							
Potentiometers																		X		
Printed-Circuit Bd											X									
Reactors																		X		
Receptacles													X							
Rectifiers																		X		
Relays																		X		
Resistors																			X	
Resolvers																			X	
Rings						X														
Rivets						X														
Screws				X		X														
Sheet Metal Work					X	X	X	X	X											
Silver Soldering								X												
Spacers						X														
Springs						X														
Spot Welding									X											
Studs				X		X														
Suppressors																		X		
Switches																		X		
Synchros																		X		
Terminals						X														
Terminal Boards					X															
Test Points						X													X	
Thermostats																		X		
Timers																		X		
Transformers																		X		
Transistors																X				
Tube Socket (Included with tube – Do not count)																				
Tubes														X	X					
Washers (Included with bolt – Do not count)																				

Fig. A3-1. (Contd.) Supplementary list for determining categories for components not listed on estimating form.

this work must be separated and given special attention, and the proper charts developed.

Note that there are often three charts for one type of work, to provide for the simple, average, or complex job. The same is done for various degrees of precision.

The charts should be read as accurately as possible for consistency. This can be accomplished by the use of a clear plastic strip about 9 in. long. Rather than use the edge, it will be found that more accurate readings can be made if a line is scribed, or etched, lengthwise on the strip, near its center. This line is used to locate all points of reference and reading points.

It is most important to become familiar with the content of each chart, as described on the particular facing page.

4. Determine the hours per standard hour at unit one, and the learning-curve percent for each type of work, by use of the charts, Figs. A3-31 through A3-39, as per the example on the charts. The estimating form provides for these to be noted in parentheses in a convenient area.

   a. Charts are given so that this information can be determined in total for a system where time is limited or greater accuracy is not required.

   There are also charts for inspection and test, mechanical, sheet metal, and electrical, so that the total will be properly weighted for the particular mix or percentage of each type of work.

   Note that the charts for hours per standard hour at unit one have two graduated elements that control the unit amount. These are *newness* and *complexity*. (See Figs. A3-28, A3-31, A3-33, A3-36, and A3-38.)

   The scale on complexity is self-explanatory for the most part, but newness can be a problem at times. The following directions are helpful:

   The zero (0) scale for newness should be used only after an initial production quantity, or a preproduction program of 25 items or more. A small quantity of prototypes should not be considered as a preproduction program. Hence, the production will be a continuation of the prototypes with the same unit one amount until hard tools are put into use. These charts must be given careful consideration.

   The second portion of this item is to determine the *learning-curve percent*. This is done by applying the unit one hours per standard hour and the maximum scheduled amount per month (delivery rate) to the learning-curve percentage charts (Figs. A3-30, A3-32, A3-34, A3-35, A3-36, and A3-39.) The operation is simple and direct.

5. Determine the hours per standard hour for the program by using the chart (Fig. A3-40).

   a. This is a routine operation that can be and usually is accomplished by the use of learning-curve tables. For extremely accurate work this is the best method, but for rapid work the procedure of finding the midpoint of the program on the unit curve and multiplying it by the hours per standard hour at unit one, has been developed into a chart-reading operation that is sufficiently accurate for most work. The example as shown in Fig. A3-40 provides the routine. This example is for a follow-on program. An initial program requires that the midpoint be located by the "index for unit one of an initial program," and the last unit as located on the dashed line. After the midpoint is located, the process is the same as for a follow-on quantity.

6. Multiply the calculated standard hours for the program for each type of work, by the hours per standard hour for the respective types of work. These can then be totaled to provide the total estimated hours for the program.

Estimating Forms

Two types of estimating forms have been developed.

1. One, on 8½ X 11 vellum, will accommodate the labor only for one program.
2. The second is on 11 X 17 vellum, and will handle both a preproduction and a production program, or two production quantities, on the one sheet. There is also a place for the material summary, and at the bottom of the sheet the hours and material can be extended to the price level. This provides for the complete estimate on one sheet.

Example. A typical black-box has been estimated, using the sample estimating forms. Figure A3-2, the work sheet, has been used to tabulate the elements and components, and also to price the material. (See also the instructions for Fig. A3-11, Chart 8.)

Figure A3-3 is a minimum estimating form that is used to receive the totals as tabulated on Fig. A3-2. From these tabulated amounts, the standard hours are determined from the respective charts having numbers that correspond to the lines of the estimating form. Each chart that was used has a dashed line giving the reading for this example.

The next step is to determine the hours per standard hour. This can be worked out by following the examples on the charts (Figs. A3-31 to A3-39), and reading the pertinent text.

Figure A3-4 is an alternative estimating form that has several advantages over Fig. A3-3 when estimating alternative quantities for a program, or when both a minimum and a probable estimate are required. It also has provision for summarizing the material costs and extending the hours and material costs to the price level. The same black box is used for an example on this form, showing its possibilities.

METHOD NO. 5 WORK SHEET

Project: EXAMPLE
Unit No. 987654-100
Unit Name: BLACK-BOX
Date: .... Ref No: .... Sheet 1 of 1

Indentured Item Description	Qty Req'd	Kind or Description	Size	Unit Cost	Purchased Parts	Sub-Contract	Other	Total
MECHANICAL								
CASTINGS	10	ALUMINUM	.5#			$100.00		$100.00
GEARS	24	STEEL	1.#	.30			7.20	7.20
MACHINED PTS	75	STEEL	1.#				8.00	8.00
BOLTS	24			.15	3.60			3.60
SCREWS	50			.10	5.00			5.00
SHEET METAL								
SHT.MTL	60	ALUMINUM	.4#				15.00	15.00
SCREWS	90			.08	7.20			7.20
RIVETS	240				0.80			0.80
TERMINALS	150			.10	15.00			15.00
WELDING								
ELECTRICAL								
PR.CIR.BDS	40			0.90			36.00	36.00
HARNESSES	1			0.08			60.00	60.00
CONNECTORS	6			5.00	30.00			30.00
PLUGS	40			3.00	120.00			120.00
INDICATOR LIGHTS	10			3.00	30.00			30.00
TRANSISTORS	80	SILICON		6.00	480.00			480.00
DIODES	322			0.50	161.00			161.00
TRANSFORMERS	8			20.00	160.00			160.00
COILS	10			4.00	40.00			40.00
CAPACITORS	250			1.00	250.00			250.00
RESISTORS	300			0.40	120.00			120.00
(SUBTOTAL) FACTOR %								
Totals					1422.60	100.00	126.20	1648.80

Legend:

Q = Quantity
# = Lbs. of Row Mat.
θ = Average Diam.

P = Number of Parts
□ = Lbs. of Row Mat.
S = Quantity of Spots
In. = Inches of Weld
Cu. In. = Cubic In. Part Size

□" = Square Inches - Average Size
W = Quantity of Wires
T = Number of Terminals

Estimator: ....
Remarks: -

Fig. A3-2. Work sheet of a typical black-box based on sample estimating forms.

Unit No. 987654-100   Name BLACK-BOX ....   Ref. No. ....   Date ....
Size 10×12×20..... Wt 30 Lb   Similar ⎱(No .... ⊞ ⊟ ......%   Engineer ....   Phone ....
PROGRAM: Date – Start .... to Unit ⎰(Name   Estimator ....
End ....   From Unit 5.... through Unit 50...........

CHART NUMBER / ITEM	Min	Av	Max	VALUE 1 Am't.-Unit	VALUE 2 Am't.-Unit	Fab.	Ass'y	Test	F.	A.	T.	Fab.	Ass'y	Test	TOTAL HOURS
				*	*	STANDARD HR			HR / STD. HR			PROGRAM ESTIMATE			
MECHANICAL															
1. Castings				10 Parts	20 Lb.	17.3	1.8					(T₁8.5)	(T₁8.2)	(T₁8.1)	
2. Gears		✓		24 Parts	1.5"Av. ⊖	18.5	5.0								
3. Machined Parts		✓		75 Parts	8 Lb.	18.9	9.5					(C%88)	(C%903)	(C%96)	
4. Bolts; Screws; Etc.				74 Qty.	10 Q†Cstg.	3.2	1.2								
5. ††															
Total – Mechanical						57.9	17.5		3.9	4.4		225.8	77.0		302.8
6. Inspect & Test Assemblies								2.1			6.2			13.0	13.0
SHEET METAL															
7. Boxes; Racks; Chassis; Etc.				60 Parts	15 Sq.Ft.	4.0	3.4								
8. Screws; Rivets; Terminals				279 Parts	60 S.M. Pts	2.4	5.4					(T₁ 11.2	11.2)	(T₁8.1)	
9. Welding – Heliarc – Hand				10 Parts	20 Lin. In.		0.3								
10. Brazing; Silver Soldering				Parts	Cu. In.							(C% 91.5	91.5)	(C%96)	
11. Spot Welding				16 Parts	150 Welds		0.5								
12.															
Total – Sheet Metal						6.4	9.6		6.3	6.3		40.3	60.5		100.8
13. Inspect & Test Assemblies								0.7			6.2			4.3	4.3
ELECTRICAL															
14. Circuit Welding				Welds	--										
15. Printed-Circuit Boards		2s		40 Qty.	30 A.S.I.	36.0									
16. Harnesses		✓		450 Wires	20"Length		28.5								
17. Plugs – All Types				46 Qty.		0.6	2.4								
18. Tubes With Sockets				10 Qty.	5 Q.Chas.	0.2	0.2								
19. Miniature Tubes				Qty.	A. Term.										
20. Transistors				80 Qty.	--		2.3								
21. Diodes				322 Qty.	--		5.9								
22. Multi-Term. Components				18 Qty.	5 A. Term.		1.0					(T₁ 11.2	28.2)	(T₁11.8)	
23. Pig-Tail Components				550 Qty.	--		11.9								
24.															
25.												(C%91.5	82.6)	(C%93.7)	
Total – Electrical						36.8	52.2		6.3	8.9		231.8	464.6		696.4
26. Inspect & Test Assemblies								7.8			7.6			59.3	59.3
SUBTOTAL						101.1	79.3	10.6							
27. System Test								8.7			7.6	(T₁11.8; C%93.7)		66.1	66.1
GRAND TOTAL						101.1	79.3	19.3				497.9	602.1	142.7	1,242.7

* Two elements are used to determine hours on most charts.
† Or machined parts receiving bolt holes and bolts.
†† Chart numbers 5, 12; and 25 have not been used in our example, but are provided to allow for growth.

(M) Mechanical Standard Hours
(E) Electrical Assembly & Test Standard Hours (Quality Control)

Fig. A3-3. Minimum estimating form used to receive the totals tabulated on A3-2.

Size: 10 × 12 × 20 .... Similar to: .... (No. ....)

Wt: 30 .... Lb .... (Name ....)

Engineer: .... Phone .... Estimator ....

Alternative Qty. 100 Ea. @ 5/Mo. PROGRAM

46 Ea. @ 5/Mo. PROGRAM

Start Date .... End Date ....

From Unit No. 5 .... Through Unit No. 50

HR/STD. H. PROGRAM ESTIMATE

Start Date .... End Date ....

From Unit No. 5 .... Through Unit No. 100

HR/STD. H. PROGRAM ESTIMATE

Date: ....

MATERIAL ESTIMATE

CHART NUMBER ITEM	M/V	A/q	M/n	VALUE 1 Am't. – Unit	VALUE 2 Am't. – Unit	STANDARD HR Fab.	Ass'y Test	HR/STD. H. PROGRAM ESTIMATE F.	A.	T.	FAB.	ASS'Y TEST	TOTAL HOURS	F.	A.	T.	FAB.	ASS'Y TEST	TOTAL HOURS	COST EACH	TOTAL COST
MECHANICAL:																					
1. Castings				10 Parts	20 Lb	17.3	1.8				8.2	8.1					8.5	8.2 8.1			
2. Gears	✓			24 Parts	1.5" Av ⊖	18.5	5.0				90.3	96					88	90.3 96			
3. Machined Parts	✓			75 Parts	8 Lb	18.9	9.5														
4. Bolts; Screws; Etc.				74 Parts	10 Cstg.* Q.	3.2	1.2														
5.																					
Total – Mechanical						57.9	17.5	3.9	4.4		225.8	77.0	302.8	3.4	3.8		196.9	66.5	263.4		
6. Inspect & Test Assemblies							2.1		6.2				13.0		5.9			12.4	12.4		
SHEET METAL:																					
7. Boxes; Racks; Chassis; Etc.				60 Parts	15 Sq. Ft.	4.0	3.4														
8. Screws; Rivets; Terminals				279 Parts	60 S.M.Pts	2.4	5.4				11.2	8.1					11.2	11.2 8.1			
9. Welding – Heliarc Hand				10 Parts	20 In.		0.3				96	96					96	96			
10. Brazing; Silver Soldering					Cu. In.		0.5				91.5	96					91.5	91.5 96			
11. Spot Welding				16 Parts	150 Welds																
12.																					
Total – Sheet Metal						6.4	9.6	6.3	6.3		40.3	60.5	100.8	5.8	5.8		37.1	55.7	92.8		
13. Inspect & Test Assemblies							0.7		6.2				4.3		5.9			4.1	4.1		
ELECTRICAL:																					
14. Circuit Welding	✓			40 Qty.	30 ** Welds	36.0															
15. Printed-Circuit Boards				450 Wires	20 Length		28.5														
16. Harnesses				46 Qty.		0.6	2.4														
17. Plugs – All Types				10 Qty.	5 Q.Chass	0.2	0.2														
18. Tubes with Sockets				Qty.	A. Term.																
19. Miniature Tubes				Qty.			2.3														
20. Transistors				80 Qty.			5.9														
21. Diodes				322 Qty.			1.0				28.2	11.8					28.2	28.2 11.8			
22. Multiterm. Components				18 Qty.	5 A. Term.		11.9				82.6	93.7					82.6	82.6 93.7			
23. Pig-Tail Components				550 Qty.																	
24.																					
25.																					
Total – Electrical						36.8	52.2	6.3	8.9		231.8	464.6	696.4	5.8	6.8		213.4	355.0	568.4		
26. Inspect & Test Assemblies							7.8		7.6				59.3		7.0			54.6	54.6		
27. System Test						101.1	79.3 10.6		7.6		11.8	66.1	66.1		7.0		11.8	60.9	60.9		
SUBTOTAL						101.1	79.3 19.3														
GRAND TOTAL													1,242.7					132.0	1056.6		

	FABRICATION			ASSEMBLY											
	Hours	Rate	Labor $	Oh.@150	Hours	Rate	Labor $	Oh.@100	Prem. @1%	Subtotal	Material	M.B. @10%	Subtotal	Other	Subtotal
Each For 46	498	3.00	1494	2241	745	2.00	1490	1490	30	6745	1,731	173	8649		8649
Total For 46	22,908			34,270						79,626					
Each For 100	447	3.00	1341	2012	609	2.00	1218	1218	26	5815	1,731	173	7719		7719
Total For 100	44,700			60,900				175,100					175,100		

	Prod. Supp.	Total Cost	Pkg.	Factory Cost	G&A@2% F/P 10%	Approx. Price
	150	8799	50	8849		10,422
					(Factor)	419,412
	140	7859	40	7899	1.1778	8,303
						930,300

Mechanical Standard Hours: Total 128.2

Electrical Standard Hours: Total 71.5

G&A@2% F/P 10% .... Total 1648.80 173.00

* Or Machined Parts Receiving Bolts, Screws, etc.

** Average Board Size – Square Inches

Note: No's. 5, 12, and 25 have been left free for unusual elements or growth.

Fig. A34. Alternate estimating form (same black box) providing for varying quantities, minimum and probable estimates.

219

Standard Hour Charts
Typical Nomograms

CASTINGS: FABRICATION AND ASSEMBLY

Mechanical

The nomogram and the formulas for this item (Fig. A3-5) were developed by plotting the standard hours as applied to the planning sheets for a representative group of castings. It was found that the standards for aluminum and stainless steel could be plotted on the same charts because the machinability and the weights are equalizing elements.

The standard hours represented are for processing, necessary inspection, and all machining excluding that done to accommodate an installed part or bolt. The assembly standard hours are for the necessary handling during the assembly operations.

NOTE: In all cases, the relative nomogram faces the text: For reference also, remember that the *chart* numbers here, tie in with the numbers on the sample black box in Figs. A3-2, A3-3, and A3-4. Greater accuracy and ease of reading these charts can be realized in some cases by reading the values at ten times the quantities, and then moving the decimal point one place in the answer.

CHART 1. CASTINGS: FABRICATION AND ASSEMBLY

MECHANICAL

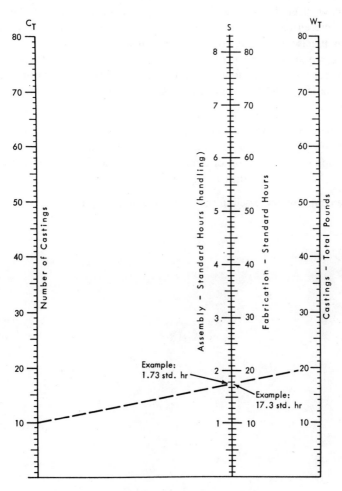

Fig. A3-5.   Fabrication: $S_F = 0.32\ C_T + 0.71\ W_T$

Assembly: $S_A = 0.032\ C_T + 0.071\ W_T$

GEARS: COMMERCIAL, PRECISION (1), PRECISION (2)

Mechanical

Charts for this item are developed by plotting the standard hours as applied to the planning sheets for a representative group of gears in each class. The type of work is represented by the precision required. This is also true of the material; the more precise the gears, the tougher and harder the material. Necessary machining required in the next assembly to receive the gears is also included. Assembly consists of the required handling and assembly operations. There are three charts (Figs. A3-6A, A3-6B, and A3-6C) provided for gears. They are, respectively:

Commercial — Average work.

Precision (1) — Maximum precision attained by many firms.

Precision (2) — Precision difficult to attain. Few firms are able to produce these.

These elements are expressed in the formulas and nomogram.

CHART 2A. GEARS: COMMERCIAL

MECHANICAL

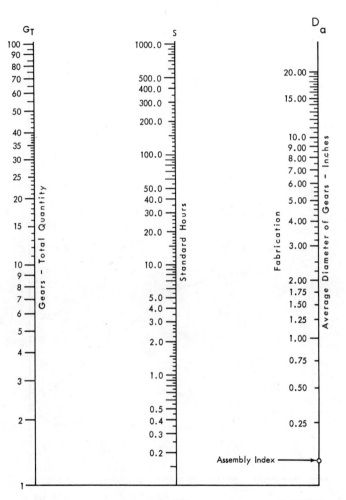

Fig. A3-6A.   Fabrication: $S_F = $ Antilog $\left\{\left[\text{Log } G_T + 0.9 \text{ (Log } D_a)^{1.036}\right] -0.33\right\}$

Assembly: $S_A = 0.13 \, G_T$

CHART 2B. GEARS: PRECISION (1)

MECHANICAL

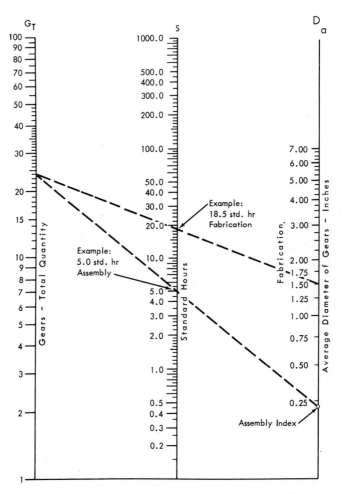

Fig. A3-6B. Fabrication: $S_F = \text{Antilog}\left\{\left[\text{Log } G_T + 0.9\,(\text{Log } D_a)^{1.07}\right] -0.26\right\}$

Assembly: $S_A = 0.215\,G_T$

CHART 2C. GEARS: PRECISION (2)

MECHANICAL

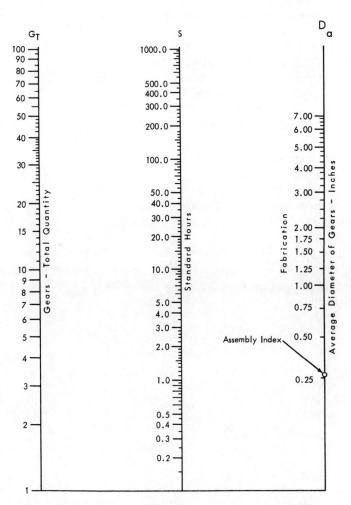

Fig. A3-6C. Fabrication: $S_F = \text{Antilog}\left\{\left[\text{Log } G_T + 0.91 \,(\text{Log } D_a)^{1.037}\right] -0.0773\right\}$

Assembly: $S_A = 0.33 \, G_T$

MACHINED PARTS: SIMPLE, AVERAGE, PRECISION

Mechanical

The nomograms and formulas are developed by plotting the standards as applied to a representative group of machined parts. (See Figs. A3-7A, A3-7B, and A3-7C.)

Note that the basic element used for a measure of value is the weight of the raw material. This can be readily estimated by the volume indicated by the maximum dimensions of the part. For most purposes, the following is sufficiently accurate for material weights per square inch and per cubic inch.

1. The weight of aluminum, and its various alloys, is relative to its thickness as one is to ten. Thus, aluminum 0.167 in. thick, weighs 0.0167 lb psi. Then, 1-in. aluminum weighs 0.1 lb psi. This is also a cubic inch.
2. Steel and brass weigh approximately three times as much as aluminum, or 1-in. thick steel weighs 0.3 lb psi (cubic inch).
3. Magnesium weighs two-thirds as much as aluminum, or 1-in. thick magnesium weighs 0.067 lb psi (cubic inch).

The fabrication hours include the time to machine the part, the time required to machine holes to receive the part, and the necessary fabrication inspection. The assembly includes the time required for handling and installing the part in the next assembly.

CHART 3A. MACHINED PARTS: SIMPLE OR COMMERCIAL

MECHANICAL

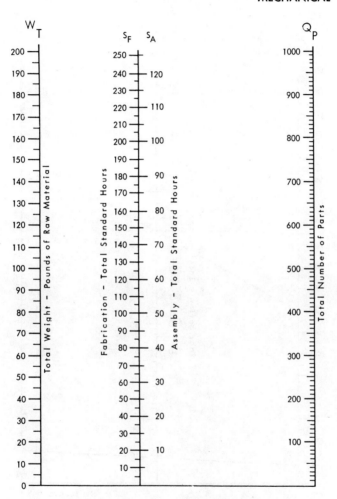

Fig. A3-7A.   Fabrication: $S_F = 0.803\, W_T + 0.0913\, Q_P$

Assembly: $S_A = 0.4015\, W_T + 0.0457\, Q_P$

CHART 3B. MACHINED PARTS: AVERAGE

MECHANICAL

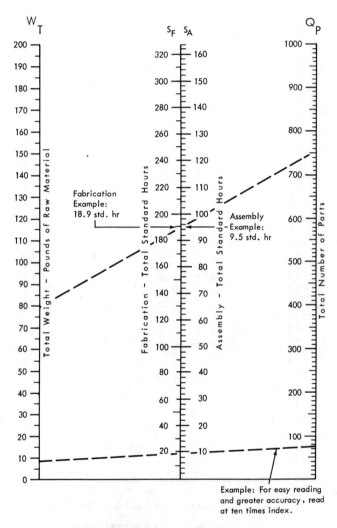

Fig. A3-7B. Fabrication: $S_F = 0.8\ W_T + 0.168\ Q_P$

Assembly: $S_A = 0.4\ W_T + 0.084\ Q_P$

CHART 3C. MACHINED PARTS: PRECISION OR COMPLEX

MECHANICAL

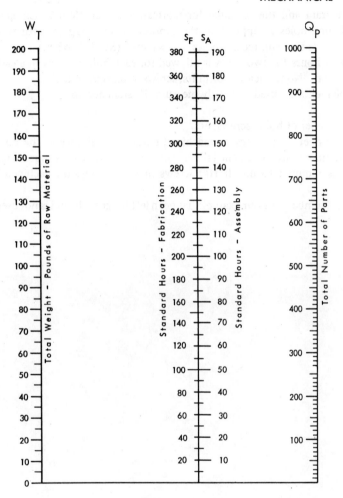

Fig. A3-7C. Fabrication: $S_F = 0.82\,W_T + 0.22\,Q_P$

Assembly: $S_A = 0.41\,W_T + 0.11\,Q_P$

BOLTS, SCREWS, ETC.

Mechanical

The nomogram and the formulas for fabrication are developed from standards for machining holes as applied to the representative group of castings, and/or machined parts that will receive bolts, screws, etc. (See Fig. A3-8.)

Standard hours for two holes is allowed for each bolt or screw in a casting or machined part. No fabrication of bolts, screws, etc., is included.

Two elements are used to jointly measure the standard hours.

1. The number of bolts, screws, etc.
2. The number of castings or machined parts that will receive the bolts, etc. (More time is involved in installing 100 screws in 50 castings or machine parts than is required to install 100 screws in two castings or machined parts.)

NOTE: Number 5 in our example has been left for growth, and is not used here.

CHART 4. BOLTS, SCREWS, ETC.

MECHANICAL

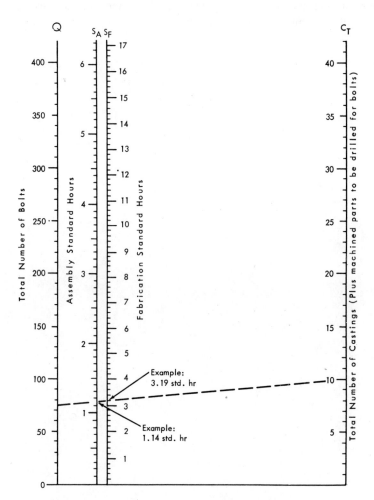

Fig. A3-8.   Fabrication: $S_F = 0.03405\, Q + 0.069\, C_T$

Assembly: $S_A = 0.013\, Q + 0.02\, C_T$

MECHANICAL INSPECTION TESTING

Mechanical

The formula and the nomogram are developed from labor standards as applied to representative assemblies. (See Fig. A3-9.)

CHART 6. MECHANICAL INSPECTION AND TESTING:

MECHANICAL

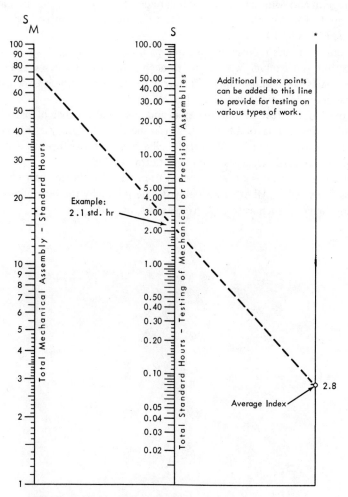

Additional index points can be added to this line to provide for testing on various types of work.

Example: 2.1 std. hr

2.8

Average Index

* This line with only one point occurs in a number of charts. Additional points can be added as the system grows.

Fig. A3-9. $S = 0.028\,S_M$

FABRICATION AND ASSEMBLY

Sheet Metal

The nomogram and formulas for this item are developed by plotting the standard hours as applied to the planning sheets for a representative group of sheet metal parts. (See Fig. A3-10.)

The standard hours represented include painting, processing, inspection, and all fabrication, but exclude work done to accommodate the screws, rivets, etc., and other parts and components. This provides the proper base to which any type of assembly technique can be applied.

The total square feet of raw material and the total number of parts, when used jointly, provide a favorable index. This is the key to accuracy in this method. It should be obvious that it will usually take less time to fabricate a sheet of metal 4′ X 10′ into 10 pieces than to form it into 1000 pieces.

The assembly portion is a token amount to provide for the necessary handling during assembly only. All other assembly is determined by the type of assembly-connecting elements such as screws, rivets, and spot-welds.

CHART 7. FABRICATION AND ASSEMBLY

SHEET METAL

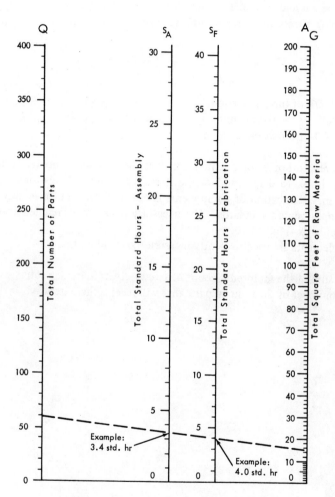

Fig. A3-10.   Fabrication: $S_F = 0.035\,Q + 0.134\,A_G$

Assembly: $S_A = 0.04\,Q + 0.0725\,A_G$

SCREWS, RIVETS, TERMINALS, ETC.

Sheet Metal

The nomogram and formulas are developed from the standard hours applied to the planning sheets for a representative group of sheet metal parts. (See Fig. A3-11.)

Instructions

1. Use 100% of the screw count.
2. Use 60% of the rivet count.
3. Use 30% of the terminal count.

The standard hours for this item includes the standard hours to fabricate the necessary holes to accommodate the items called out. It also includes their installation. Fabrication of components is not included. The assembly portion is based on accepted standards. The necessary handling is included in the sheet metal standards—Fig. A3-10.

Two elements are used to jointly measure the standard hours.

1. The sum of the factored amount of screws, rivets, terminals, etc.
2. The number of sheet metal parts that will receive these items.

CHART 8. SCREWS, RIVETS, TERMINALS, ETC.

SHEET METAL

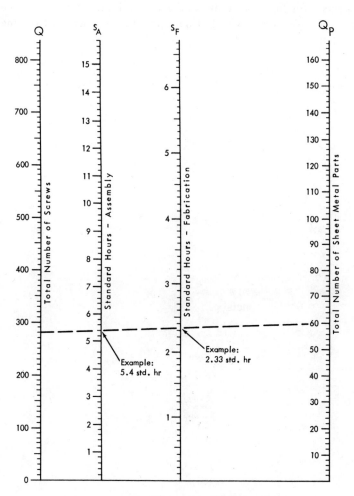

Fig. A3-11.   Fabrication: $S_F = 0.004\,Q + 0.02\,Q_P$

Assembly: $S_A = 0.015\,Q + 0.02\,Q_P$

WELDING: HELIARC

Sheet Metal

The nomogram for this work provides the standard hours for each of the three types of work. (See Fig. A3-12.) Three formulas were derived from the basic standards data.

CHART 9. WELDING: HELIARC

SHEET METAL

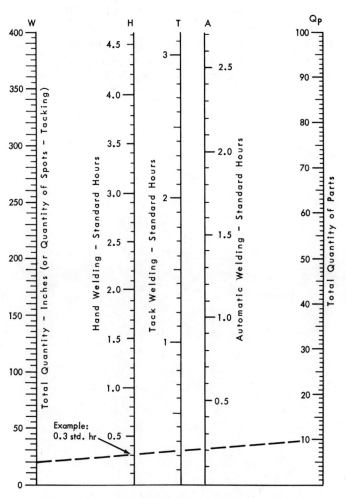

Fig. A3-12. $S_H = 0.0076\ W_I + 0.0157\ Q_P$

$S_T = 0.00393\ W_S + 0.016\ Q_P$

$S_A = 0.00275\ W_I + 0.016\ Q_P$

BRAZE, OR SILVER SOLDER

Sheet Metal

The standard hours are to file, fit, clean, and braze (or solder). (See Fig. A3-13.)

The nomogram was obtained from accepted standards. The elements as given for various parts were tabulated and charted, size vs. standard hours, to form the basis of the final chart. It is basically standard hours per cubic inch, but the standard hours per inch varies with the size of the assembly.

The handling of the parts is included in Fig. A3-10.

CHART 10. BRAZE, OR SILVER SOLDER

SHEET METAL

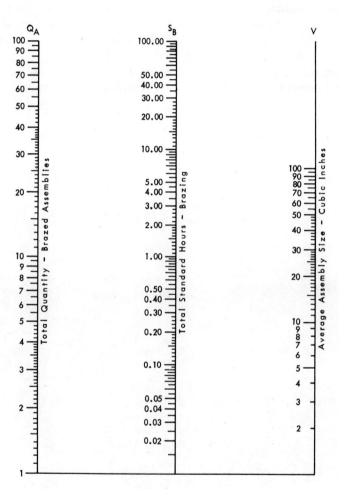

Fig. A3-13. $S_B = \text{Antilog} \left[(\text{Log } Q_A + 0.7103 \text{ Log } V) - 2 \right]$

SPOT WELDING

Sheet Metal

This nomogram provides the total standard hours for spot welding. The standard hours are determined from the total number of parts to be welded and the total number of welds. Two scales are given. One is for items under 500 cu in. in size, and the other is for items over 500 cu in. (See Fig. A3-14.)

Accepted standards are used as the base for the formulas and the nomogram. NOTE: Number 12 has been left for growth, and is not used here.

CHART 11. SPOT WELDING

SHEET METAL

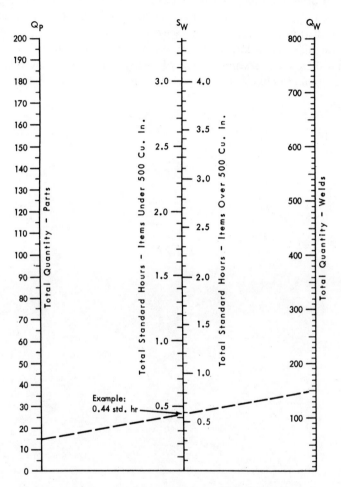

Fig. A3-14. Items under 500 cu. in.: $S_W = 0.008\ Q_P + 0.00215\ Q_W$

Items over 500 cu. in.: $S_W = 0.011\ Q_P + 0.00285\ Q_W$

INSPECTION AND TEST: SHEET METAL ASSEMBLIES

Sheet Metal

Inspection is often the only element required for sheet metal. The nomogram is therefore made for inspection or testing as individual items. It also provides for a total of both inspection and testing. (See Fig. A3-15.)

The standard hours for inspection and testing are based on their being relative to the total sheet metal assembly standard hours. This relationship is based on percentages found to represent the experienced average.

CHART 13. INSPECTION AND TEST: SHEET METAL ASSEMBLIES

SHEET METAL

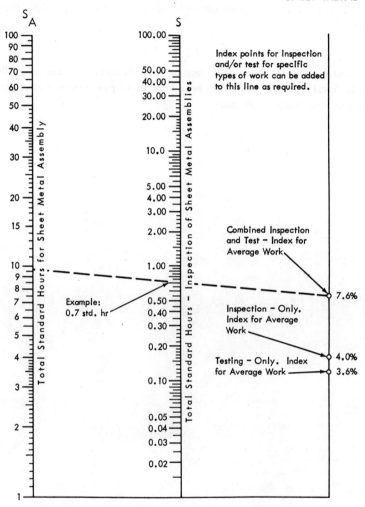

Fig. A3-15.   Inspection $S_I = 0.040\, S_A$

Testing $S_T = 0.036\, S_A$

Insp. & Test $S_{I-T} = 0.076\, S_A$

CIRCUIT WELDING: ELECTRIC SPOT WELDER

Electrical*

The standard for this nomogram has been taken at 800 welds per day. (See Fig. A3-16.)

---

*Throughout this book the general term *electrical* has been used in all discussion and illustrations. It should be understood that its use is in all cases synonymous with *electronic.* No specific differentiation has been made.

CHART 14. CIRCUIT WELDING: ELECTRIC SPOT WELDER

ELECTRICAL

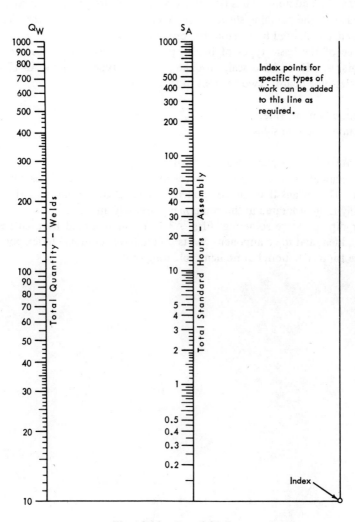

Index points for specific types of work can be added to this line as required.

Fig. A3-16. $S_A = 0.01\ Q_W$

PRINTED-CIRCUIT BOARDS (1), EYELET-TYPE

Electrical

The formulas and nomograms (see also Fig. A3-17B) are based on standard hours as applied to the planning sheets of representative groups of boards. These are tabulated and charted to develop the formulas.

Three of the basic types of boards presently in use are represented on two nomograms. These have scales for the respective types. Figure A3-17A is for the eyelet-type board. Two versions of this are in use:

1. Printed on one side.
2. Printed on both sides.

Figure A3-17B is for plated-through boards.

The standard hours are for all processing and machining. No assembly is included. This means that the components are installed by hand soldering and all assembly is incorporated in the component assembly amounts.

For dip, or wave soldering, Figs. A3-17A and B should have an assembly index added, and the component charts should have additional index points that provide for installation, but no hand soldering time.

CHART 15A. PRINTED-CIRCUIT BOARDS, EYELET-TYPE

ELECTRICAL

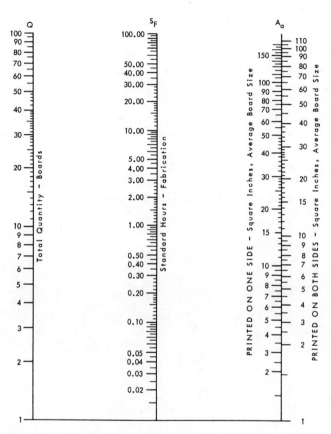

Fig. A3-17A.

Printed One Side $S_F = \text{Antilog} \left\{ \left[\text{Log } Q + 0.858 \, (\text{Log } A_a)^{1.083} \right] - 2.062 \right\}$

Printed Both Sides $S_F = \text{Antilog} \left\{ \left[\text{Log } Q + 0.7124 \, (\text{Log } A_a)^{1.218} \right] - 1.765 \right\}$

PRINTED-CIRCUIT BOARDS (2), PLATED-THROUGH

Electrical

The formulas and nomograms (see also Fig. A3-17A) are based on standard hours as applied to the planning sheets of representative groups of boards. These are tabulated and charted to develop the formulas.

Three of the basic types of boards presently in use are represented on two nomograms. These have scales for the respective types.

The eyelet-type board was covered in Fig. A3-17A. The nomogram opposite (Fig. A3-17B) is for plated-through boards.

The standard hours are for all processing and machining. No assembly is included. This means that the components are installed by hand soldering and all assembly is incorporated in the component assembly amounts.

For dip, or wave soldering, Figs. A3-17A and B should have an assembly index added, and the component charts should have additional index points that provide for installation, but no hand soldering time.

CHART 15B. PRINTED-CIRCUIT BOARDS. PLATED-THROUGH

ELECTRICAL

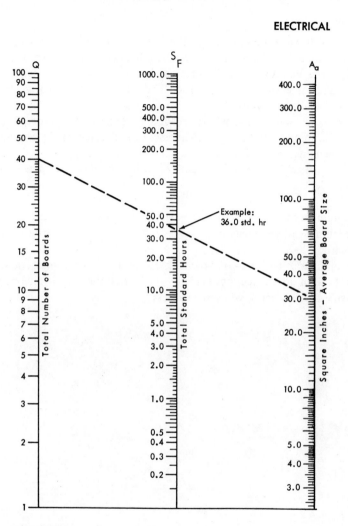

Fig. A3-17B.

Plated Through $S_F = $ Antilog $(\text{Log } Q + 0.8572 \text{ Log } A_a - 1.28)$

HARNESSES: SIMPLE AND AVERAGE

Electrical

The standard hours as determined by the formulas and nomograms (see also Fig. A3-18B, complex) include all the standard hours for wire prep and harness assembly. (See Fig. A3-18A.) The standard hours also include the standard hours to install and solder-in the harness. The lengths used are those that approximate the greatest number of wires.

The formulas and nomograms are made by tabulating the standard hours as applied to a representative group of harnesses of each category. Assembly of the plugs to the harnesses is included.

This is a critical item, and the quantity of wires should be given careful attention. Consider:

1. The number of plugs times the average number of pins per plug.
2. The number of terminals—or lugs—on tube sockets and multiterminal components.
3. The number of terminals on boards that require a harness connection.
4. The total of all connections made by the harness must be divided by two to provide the number of wires in a harness.

CHART 16A. HARNESSES: SIMPLE AND AVERAGE

ELECTRICAL

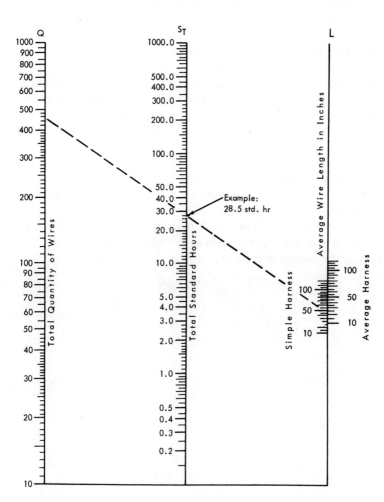

Fig. A3-18A.

Simple Harness	S_T = Antilog (Log Q + 0.191 Log L – 1.495)
Average Harness	S_T = Antilog (Log Q + 0.239 Log L – 1.509)

HARNESSES: COMPLEX

Electrical

The standard hours as determined by the formulas and nomograms (see also Fig. A3-18A) include all the standard hours for wire prep and harness assembly. (See Fig. A3-18B.) The standard hours also include the standard hours to install and solder-in the harness. Lengths used are those that approximate the greatest number of wires.

The formulas and nomograms are made by tabulating the standard hours as applied to a representative group of harnesses of each category. Assembly of the plugs to the harnesses is included.

This is a critical item, and the quantity of wires should be given careful attention. Consider:

1. The number of plugs times the average number of pins per plug.
2. The number of terminals—or lugs—on tube sockets and multiterminal components.
3. The number of terminals on boards that require a harness connection.
4. The total of all connections made by the harness must be divided by two, to provide the number of wires in a harness.

CHART 16B. HARNESSES: COMPLEX

ELECTRICAL

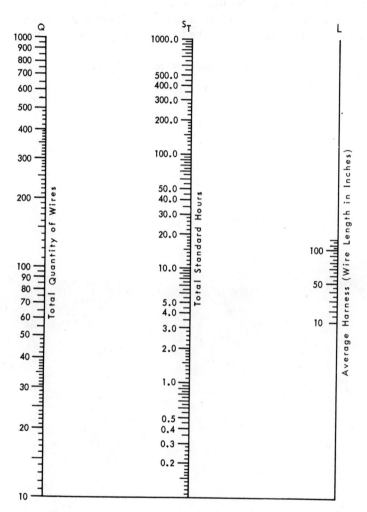

Fig. A3-18B.

Complex Harness S_T = Antilog (Log Q + 0.318 Log L - 1.55)

PLUGS: ALL TYPES

Electrical

The formulas and nomogram for this item are based on accepted standards. (See Fig. A3-19.) These basic standards are used with the assembly soldering standards which are in the standard hours for the harness as determined by Figs. A3-18A and B.

The punching of the necessary holes to receive the plugs is included in the fabrication element.

CHART 17. PLUGS ALL TYPES

ELECTRICAL

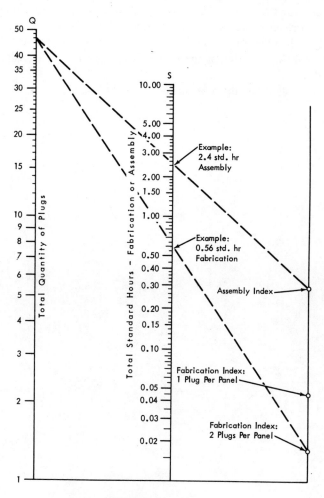

Fig. A3-19. Fabrication:

1 plug per panel $S_F = 0.0226\,Q$

2 plugs per panel $S_F = 0.0132\,Q$

Assembly: $S_A = 0.0526\,Q$

TUBES WITH SOCKETS

Electrical

The fabrication portion of this nomogram provides for the punching of the required holes for the socket. (See Fig. A3-20.) This is based on the charting of a representative group of parts as developed for Fig. A3-10.

Assembly is handling only; the riveting of the socket is provided for under Fig. A3-11, and the soldering of the terminals to the circuit is included in Figs. A3-18A and B.

Accepted standards are the base for the assembly formula.

CHART 18. TUBES WITH SOCKETS

ELECTRICAL

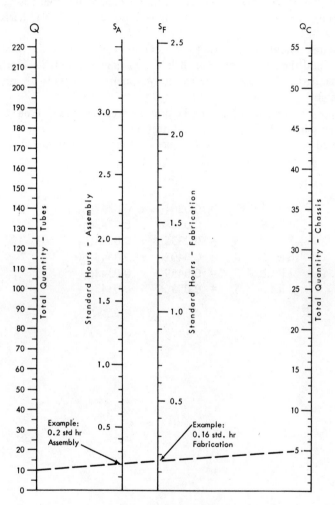

Fig. A3-20.   Fabrication: $S_F = 0.0063\ Q + 0.02\ Q_C$

Assembly: $S_A = 0.011\ Q + 0.02\ Q_C$

MINIATURE TUBES

Electrical

This nomogram may appear to deviate from the consistency that has been deliberately built into the previous chart. It is, however, six simple charts in one. (See Fig. A3-21.)

The standard hours are relative to the total number of tubes, the total number of boards, and the average number of leads per tube. There is a centerline scale in the nomogram for each of the respective quantities of leads per tube, three through eight.

Drilling of the required holes in the printed circuit boards (if used) to receive the leads, is part of the circuit board fabrication.

Accepted standards are used as the base for the nomograms and the following formulas:

Tubes with:

$$
\begin{aligned}
\text{3 leads} \quad S_A &= 0.0201Q + 0.0343Q_C \\
\text{4 leads} \quad S_A &= 0.0268Q + 0.0343Q_C \\
\text{5 leads} \quad S_A &= 0.0335Q + 0.0343Q_C \\
\text{6 leads} \quad S_A &= 0.0402Q + 0.0343Q_C \\
\text{7 leads} \quad S_A &= 0.0469Q + 0.0343Q_C \\
\text{8 leads} \quad S_A &= 0.0536Q + 0.0343Q_C
\end{aligned}
$$

CHART 19. MINIATURE TUBES

ELECTRICAL

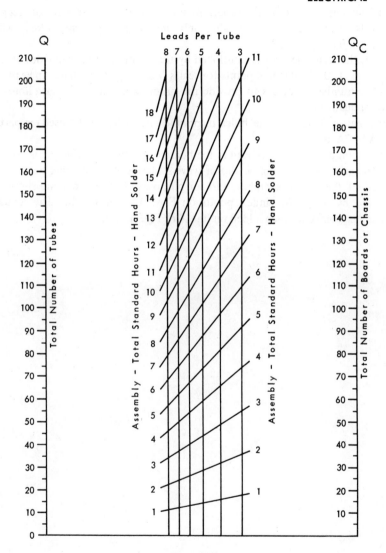

Fig. A3-21.

TRANSISTORS: HAND SOLDER

Electrical

The nomogram is based on the variations and standard hour elements of accepted standards. It includes assembly labor only: inserting the part and soldering it in by hand. (See Fig. A3-22.)

Here, the basic element is multiplied by the quantity. For quantities of transistors over 100, the decimal points for scales N and S_A of the chart can be moved the same number of places to provide the proper amount of standard hours for the total quantity.

Fabrication labor for the holes in printed circuit boards (if used) is part of Figs. A3-17A and B.

When components are inserted in the board only, and the soldering is done by an automatic method, such as dip or wave-soldering, the soldering becomes an element of the board assembly, and for purposes of this nomogram should be estimated by additional index points on Figs. A3-17A and B. The required extra index points are then added on line X of Fig. A3-22.

CHART 20.   TRANSISTORS: HAND SOLDER

ELECTRICAL

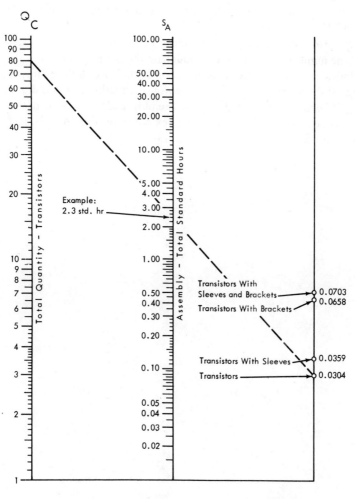

Fig. A3-22.   Transistors $S_A = 0.0304\,Q_C$

With Sleeves $S_A = 0.0359\,Q_C$

With Brkts. $S_A = 0.0658\,Q_C$

With Slvs. & Brkts. $S_A = 0.0703\,Q_C$

DIODES: HAND SOLDER

Electrical

This nomogram is based on accepted standards, and is for assembly labor only, as the fabrication (drilling of holes in printed-circuit boards, if required) is part of Figs. A3-17A and B.

For quantities of diodes over 100, move the decimal points of scales N and S_A the same number of places to accommodate the larger quantities.

The index given on line X is for insertion of the part on the board, and hand soldering. If the soldering is done by an automatic method, such as dip, or wave-soldering, it becomes a part of the board assembly, and for purposes of this nomogram should be estimated by additional index points on Figs. A3-17A and B, and the final points included on line X of Fig. A3-23.

CHART 21. SIODES: HAND SOLDER

ELECTRICAL

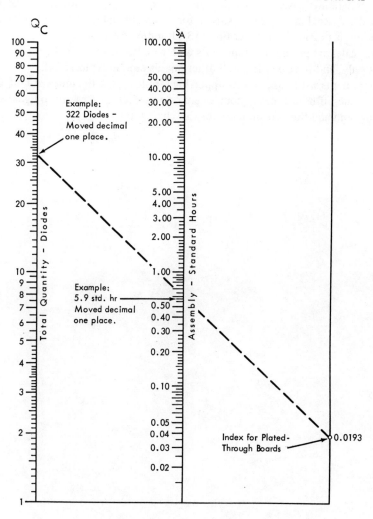

Fig. A3-23. $S_A = 0.0193\,Q_C$

MULTITERMINAL COMPONENTS

Electrical

The nomogram is based on charting the standard hour elements from accepted standards. (See Fig. A3-24.) This is for assembly only, and hand soldering; fabrication is provided for under Figs. A3-11 and/or A3-17A and B.

Two scales are given for components with wire leads: one is for high-reliability boards only, and the other is for all other boards used in typical work.

Components with lugs, or self-supporting terminals, use the index at unit 1 on the Q_T scale. If both types of components are used, it is necessary to figure each assembly and add the two amounts for this chart total.

CHART 22. MINTITERMINAL COMPONENTS

ELECTRICAL

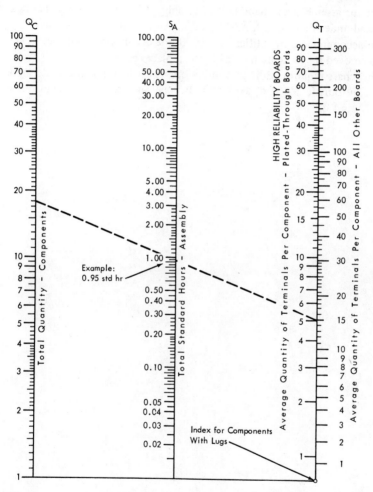

Fig. A3-24.

Plated Thru Bds.	$S_A = \text{Antilog}\left[\text{Log }Q_C + 0.889\,(\text{Log }Q_T)^{1.074} - 1.886\right]$
Other Boards	$S_A = \text{Antilog}\left[\text{Log }Q_C + 0.504\,(\text{Log }Q_T)^{1.444} - 1.913\right]$
Compnts. with Lugs	$S_A = 0.1\,Q_C$

PIG-TAIL COMPONENTS

Electrical

The nomogram is based on the standard hour elements of accepted standards, and is for assembly and hand soldering only. (See Fig. A3-25.) Fabrication is included under Figs. A3-10, A3-11, and/or A3-17A and B.

To accommodate quantities over 1000, the decimal points on scales Q and S_A can be moved the same number of decimal places as required.

This chart requires additional index points on line X if the soldering is done by an automatic method, as it will then be an assembly element. See Figs. A3-17A and B.

CHART 23.   PIG-TAIL COMPONENTS

ELECTRICAL

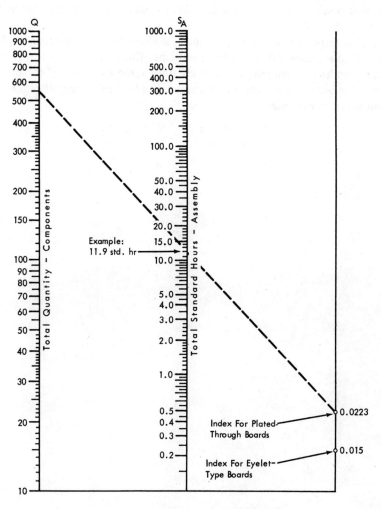

Fig. A3-25.   Eyelet Type $S_A = 0.015\,Q$

Plated Thru $S_A = 0.0223\,Q$

INSPECTION AND TEST, ELECTRICAL ASSEMBLIES

Electrical

Standards as applied to various types of units are tabulated to develop this chart. It seemed best to give the name as an index to determine the class of work. (See Fig. A3-26.)

Total of inspection and test—combined: Additional index points can be placed on line X to accommodate extra types of items. However, to promote reasonable accuracy without developing an unwieldy system, the estimating of the standard hours for inspection and testing is combined; and for further simplification these formulas can be expressed:

$$
\begin{array}{lll}
\text{Simple Units} & S_I & = \ 0.06 S_A \\
\text{Typical Units} & S_I & = \ 0.09 S_A \\
\text{Complex Units} & S_I & = \ 0.19 S_A
\end{array}
$$

CHART 26. INSPECTION AND TEST — ELECTRICAL ASSEMBLIES — TOTAL

ELECTRICAL

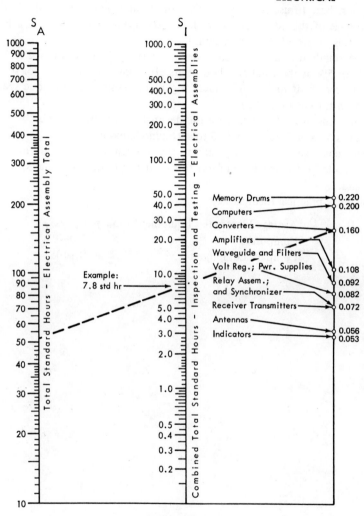

Fig. A3-26.

SYSTEM TESTING

Electrical

The best base to use for Fig. A3-27 is the total production standard hours. A percentage of this total—based on the experienced percentage—for each type of project, is suggested in the nomogram.

The standard hours for system test on radar systems is assumed to be equal to 4% or 5% of the total standard hours. It is possible, however, for the system test on certain items, such as inertial guidance systems and unique specialized equipment, to be much higher.

For items over 1000 hr, move the decimal points the same number of places for scales S_T and S_S as is required to accommodate the total.

This chart suggests an approach to the problem only, and any index points used should be located to conform to experience in the particular kind of work on hand.

Typical, practical formulas:

$$\text{Average Electronics Equipment} \quad S_S = 0.046S_T$$
$$\text{Computers} \quad S_S = 0.080S_T$$
$$\text{Complex Equipment} \quad S_S = 0.140S_T$$

CHART 27. SYSTEM TESTING

ELECTRICAL

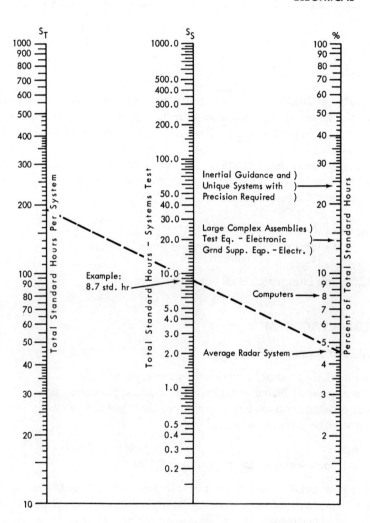

Fig. A3-27.

Learning Curves Applied to Standard Hour Estimates

In quantity production that is not geared to a machine with a specific rate of output, the labor on any unit should, up to a certain point, be less than that experienced on the preceding unit. This has been documented and well established. (See Chapter 8.)

It has also been pointed out that some years have elapsed since it was realized that the learning trend was not always on an 80% curve, as initially presumed.

The problem then arises:

a. What curve should be used?
b. What is the value that should be assigned to the first unit?

For clarity, this information is set out in step form.

PER TOTAL PROGRAM

1. Outline of Solution

a. The first thing necessary is to develop some experience records.
b. These must be evaluated and some definite conclusions developed.
c. These conclusions should be organized so that they can be readily and effectively used on proposed programs.

2. Typical Documentation for a Hypothetical Firm

There is always a limited amount of documentation on hand. Most of this is based on statistical actuals. There are also a number of factors that limit the accuracy of these records by an unknown percent. Some of these that occur during the life of the program are:

a. Changes in the outside production percent of the total program.
b. Engineering changes that increase or decrease the total labor.
c. Changes of direct labor to indirect, or vice versa.

274

This documentation provides certain limits of experience with a fairly definite pattern.

a. The extremes of the learning curve pattern for total production hours on large programs may vary from an 80% to an 88% curve, or even greater.

   i. Early programs show a constant trend—preproduction through production quantities. The same thing shows up in new, unique programs.
   ii. Retained learning shows up in later programs as a definite break between preproduction and production. This retained learning is built into the production operation through implementation and planning. As a result, production must be plotted separate from preproduction.

b. The extremes in unit 1 can best be considered by comparing the hours per standard hour at unit 1 on various programs. Bearing in mind the unknown elements that affect these records, the results are usually acceptable. The extremes for hours per standard hour at unit one are:

   i. Ten* hours per standard hour for follow-on, or programs using similar units.
   ii. Thirty* hours per standard hour for completely new, or initial programs.

3. Developing a Method of Solution

We know the limitations as experienced. We are also aware of many of the elements that affect the selection of the proper learning curve and the hours per standard hour at unit one. So the initial step is a listing of these elements and how they may possibly be applied. This is given in Table A3-1.

4. Solution

Developing a chart to determine the hours per standard hour for unit one is the first thing to accomplish, as the hours per standard hour are used in determining the slope, or percent of the learning curve.

a. Items 1 and 2 of part A of Table A3-1 are the keys to the hours per standard hour at unit 1. These are:

   i. Newness.
   ii. Complexity.

---

*These values are illustrative only and not necessarily typical for any industry.

TABLE A3-1. SELECTION OF LEARNING CURVES
FOR TOTAL PRODUCTION HOURS.

A. Elements Affecting Unit One Amount

1. Newness of Program	First element of nomogram—graduated from completely new to identical to items in production.
2. Complexity	Second element of nomogram—graduated from simple single assembly items to the extremely complex.
3. Type of Work	Department charts can be developed to amply include the type of work—it is also an element in determining complexity.
4. Hard or Soft Tools	The rate of production can be assumed to include this.

B. Elements Affecting Slope   *Action*

1. Rate of Production	Establish maximum and minimum slope for new and old.
2. Newness of Program 3. Amount of Repetitive Elements 4. Hard or Soft Tools	Included in the above by the nomogram application of the unit one element—hours per standard hour at unit one.
5. Type of Work	This is applied indirectly through the unit one element.

b. For application of the graduated values of newness and complexity it was decided to give each equal weight. Assorted data were used to develop a complexity rating based on the total standard hours per individual assembly that provided variations equal to approximately half the range being considered. This information was compiled in two addition-type nomograms, as follow, for easy use.

c. The hours per standard hour can be determined by the use of Fig. A3-28. This is one element; the other element needed to determine the learning-curve slope, or percentage, is an applicable rate of delivery scale. This can be developed as shown in Fig. A3-29. Note that the log scale at 1000 can represent the learning-curve percent or the delivery rate, or both. This is accomplished by assigning

NEWNESS: GRADUATED VALUES

Index [a]		Alternative Categories
100	*a.*	81% to 100% new units—not similar to work on hand
90	*a.*	71% to 80% new units
	b.	Similar—changes in circuitry (e.g., to transistors)
	c.	Repackage and 60% engineering change
80	*a.*	61% to 70% new units
	b.	100% new, but similar to work on hand
	c.	Repackage and 50% engineering changes
70	*a.*	51% to 60% new units
	b.	100% new, but very similar to work on hand
	c.	Repackage and 40% engineering change
60	*a.*	41% to 50% new units
	b.	Repackage and 30% engineering change
50	*a.*	31% to 40% new units
	b.	Repackage and 20% engineering change
40	*a.*	21% to 30% new units
	b.	Repackage and 10% engineering change
30	*a.*	11% to 20% new units
	b.	Repackage only
20	*a.*	6% to 10% new units
	b.	Partial repackage
10	*a.*	Up to 5% new units
0	*a.*	Reorder, after preproduction

[a] The *a, b,* and *c* descriptive elements are alternatives.

the rates of delivery to the respective curves, as indicated by experience on a representative group of projects.

This scale, and the reciprocal of the scale for hours per standard hour at unit one, provides the elements to make the nomogram (Fig. A3-30) to select the proper learning curve.

This seems to be the most consistent application of the major variables that is reasonable at this time. It is a method that can help avoid some of the hazards being encountered as a result of the extreme changes in present production program patterns.

COMPLEXITY: GRADUATED VALUES

Index [a]	Alternative Categories
100	Most Complex
	a. Inertial guidance systems
	b. Space electronics
	c. Stable platforms
90	a. Large complex assemblies—little multiple usage—many subassemblies
	b. Test equipment—electrical
	c. Ground support equipment—electrical
80	a. Complex items with much multiple usage
	b. Computers
70	a. Moderately complex items with much multiple usage
	b. Radar systems
60	a. Items above average with many duplicate and similar modules
	b. Power supplies—critical output
50	a. Average items with many duplicate and similar modules
	b. Power supplies—average
40	a. Simple items with many duplicate and similar modules
30	a. Simple electrical items with few subassemblies
	b. Racks with harnesses
20	a. Simple electrical items that are made in one simple assembly
10	a. Simple components
0	a. Parts

[a] The *a, b,* and *c* descriptive elements are alternatives. They should be used only as a pattern to develop the index for the particular work at hand.

HOURS PER STANDARD HOUR
FOR VARIOUS TYPES OF WORK

The nomographs in Figs. A3-28 and A3-30 are applicable to a system in total. This is good for some conditions, and would be adequate if all programs had the same mix of the various types of work, and if the type of work did not change.

Considering changing designs and the need to provide a system that is dynamic, it is advisable to develop charts similar to those given here, for each respective type of work. Better coverage, if desired, can be obtained by developing charts for each department, or by cost center.

The following charts, Figs. A3-31 through A3-40, are applicable to the sample estimating form developed for our hypothetical company. They provide full coverage, without going into some of the details that may be desired in actual

HOURS PER STANDARD HOUR AT UNIT 1 FOR TOTAL PRODUCTION HOURS

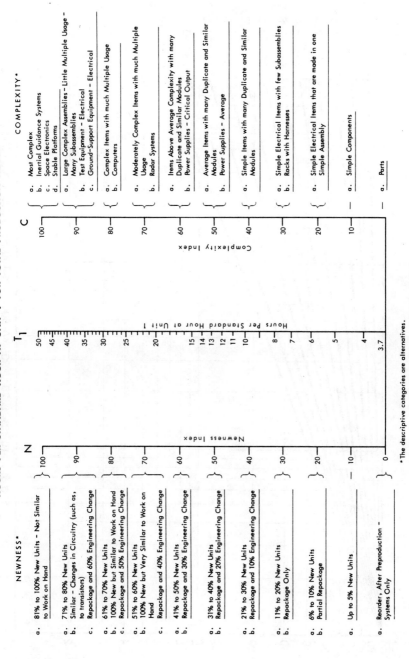

*The descriptive categories are alternatives.

Fig. A3-28. $T_1 = \text{Antilog}\,[0.005769\,(N + C) + 0.5274]$

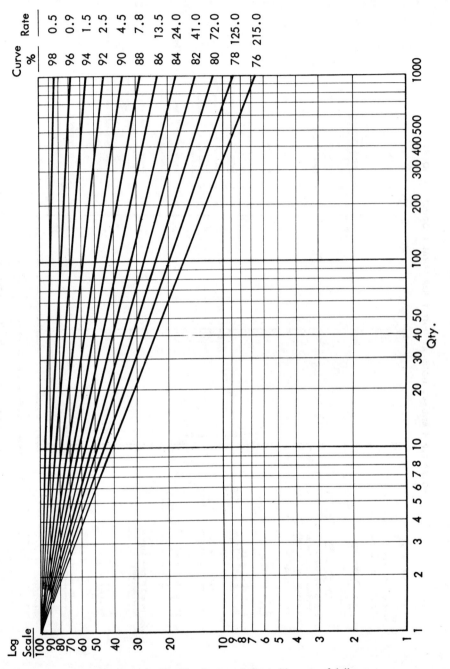

Fig. A3-29. Typical family of curves indicated by rate of delivery.

TOTAL PRODUCTION HOURS

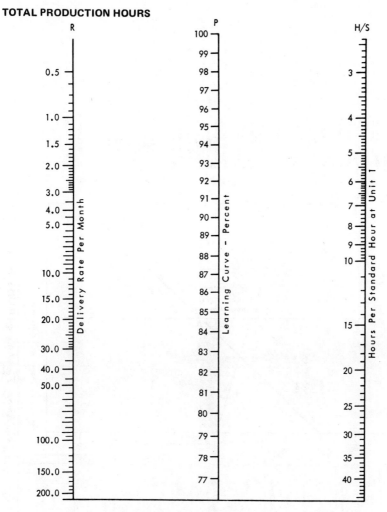

Fig. A3-30. $P = 101.3 - (4.26 \log R + 9.55 \log T_1)$

HOURS PER STANDARD HOUR AT UNIT 1 FOR TOTAL INSPECTION AND TEST

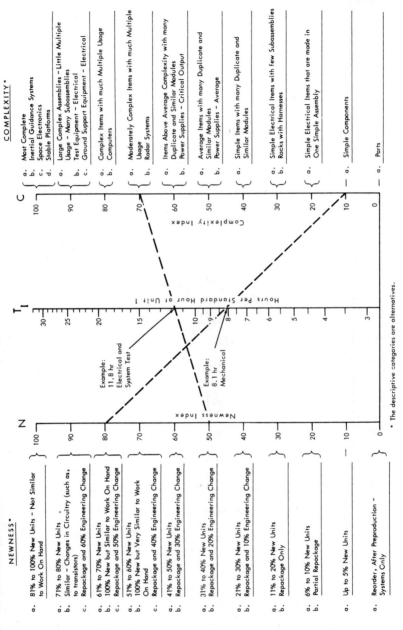

Fig. A3-31. $T_1 = \text{Antilog}\left[0.0053\,(N + C) + 0.4344\right]$

* The descriptive categories are alternatives.

INSPECTION AND TEST

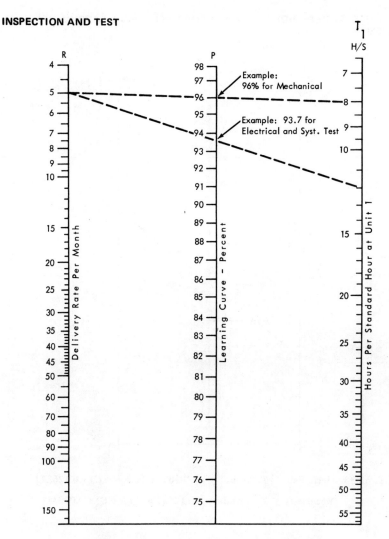

Fig. A3-32. $P = 113 - (7.7 \log R + 12.9 \log T_1)$

HOURS PER STANDARD HOUR AT UNIT 1 FOR MECHANICAL FABRICATION AND ASSEMBLY

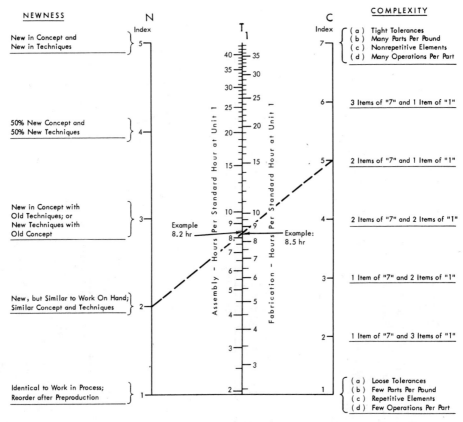

Fig. A3-33.   Mech. Fab. $T_1 = \text{Antilog} \left[0.10102 \, (1.4975 \, N + C) + 0.12684 \right]$

Mech. Assy. $T_1 = \text{Antilog} \left[0.1124 \, (1.4975 \, N + C) + 0.02048 \right]$

MECHANICAL FABRICATION

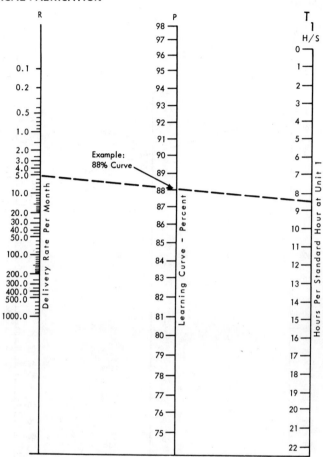

Fig. A3-34. Mech. Fab. $P = 93.4 - (1.72 \log R + 0.5 \, T_1)$

MECHANICAL ASSEMBLY

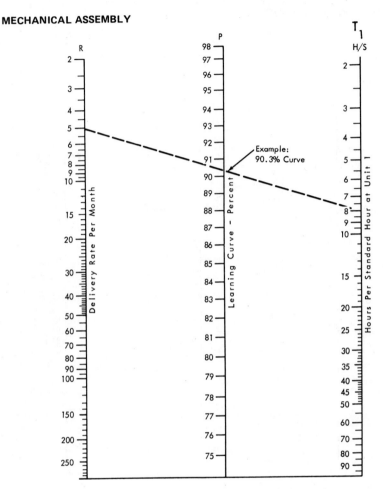

Fig. A3-35.   Mech. Assy. $P = 100.1 - (5.3476 \operatorname{Log} R + 6.7513 \operatorname{Log} T_1)$

HOURS PER STANDARD HOUR AT UNIT 1 FOR SHEET METAL

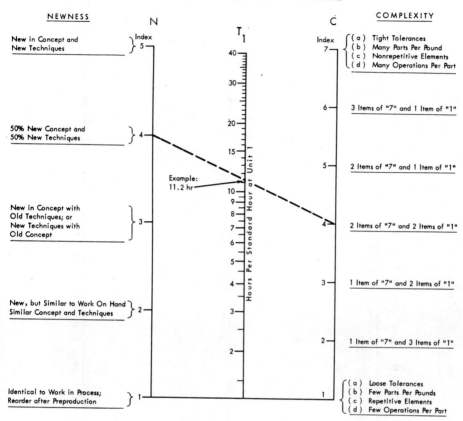

Fig. A3-36. T_1 = Antilog $(0.19\,N + 0.1274\,C - 0.2237)$

SHEET METAL

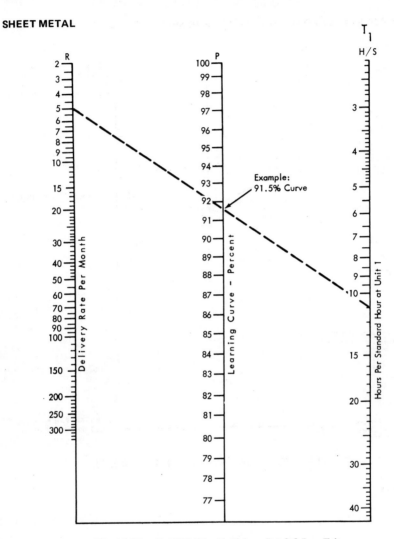

Fig. A3-37. $P = 104.27 - (4.52 \log R + 9.2 \log T_1)$

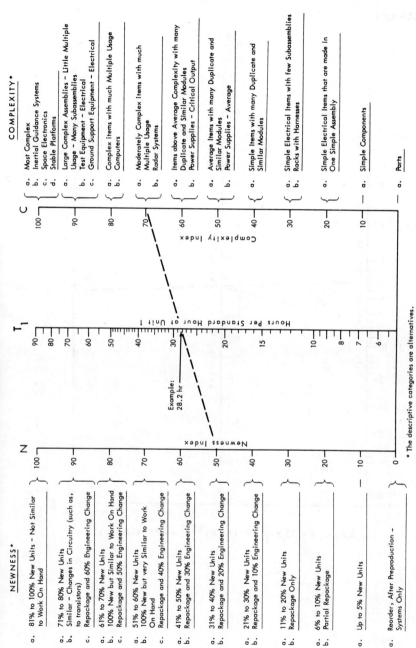

HOURS PER STANDARD HOUR AT UNIT 1 FOR TOTAL ELECTRICAL HOURS

NEWNESS*

a. 81% to 100% New Units – Not Similar to Work On Hand

a. 71% to 80% New Units Similar – Changes in Circuitry (such as, to transistors)
b.
c. Repackage and 60% Engineering Change

a. 61% to 70% New Units
b. 100% New but Similar to Work On Hand
c. Repackage and 50% Engineering Change

a. 51% to 60% New Units
b. 100% New but very Similar to Work On Hand
c. Repackage and 40% Engineering Change

a. 41% to 50% New Units
b. Repackage and 30% Engineering Change

a. 31% to 40% New Units
b. Repackage and 20% Engineering Change

a. 21% to 30% New Units
b. Repackage and 10% Engineering Change

a. 11% to 20% New Units
b. Repackage Only

a. 6% to 10% New Units
b. Partial Repackage

a. Up to 5% New Units

a. Reorder, After Preproduction – Systems Only

N

100
90
80
70
60
50
40
30
20
10
0

Newness Index

COMPLEXITY*

a. Most Complex
b. Inertial Guidance Systems
c. Space Electronics
d. Stable Platforms

a. Large Complex Assemblies – Little Multiple Usage – Many Subassemblies
b. Test Equipment – Electrical
c. Ground Support Equipment – Electrical

a. Complex Items with much Multiple Usage
b. Computers

a. Moderately Complex Items with much Multiple Usage
b. Radar Systems

a. Items above Average Complexity with many Duplicate and Similar Modules
b. Power Supplies – Critical Output

a. Average Items with many Duplicate and Similar Modules
b. Power Supplies – Average

a. Simple Items with many Duplicate and Similar Modules

a. Simple Electrical Items with few Subassemblies
b. Racks with Harnesses

a. Simple Electrical Items that are made in One Simple Assembly

a. Simple Components

a. Parts

C

100
90
80
70
60
50
40
30
20
10

Complexity Index

T_1

90
80
70
60
50
40
30
20
15
10
8
7
6

Hours Per Standard Hour at Unit 1

Example: 28.2 hr

* The descriptive categories are alternatives.

Fig. A3-38. $T_1 = \text{Antilog} \left[0.709 + 0.00618 \, (N + C) \right]$

ELECTRICAL

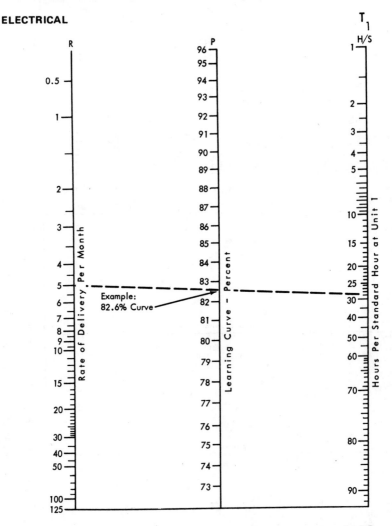

Fig. A3-39. $P = 93.8 - \left[6.24\ (\text{Log } R)^{0.815} + 4.455\ (\text{Log } T_1)^{1.463}\right]$

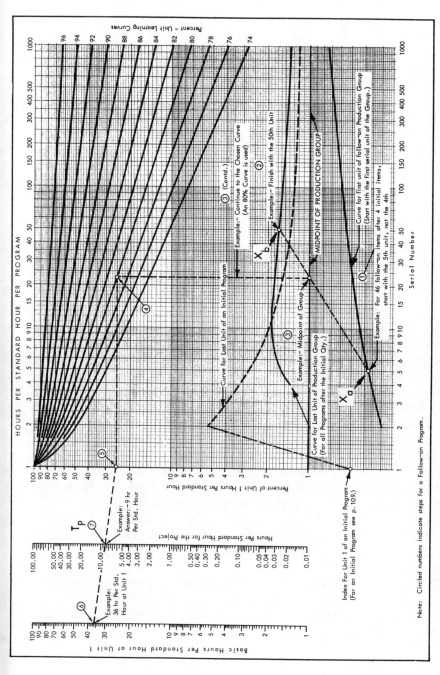

Fig. A3-40. Hours per standard hour for a program – Chart Calculator

$$T_P = \left\{ \left[X_b^L - (X_a - 1)^L \right] \div Q \right\} \cdot T_1$$

work of a specific nature. Additional charts can be developed as described earlier in this chapter.

It is highly recommended that two readings be made for each type of work. One should be for the minimum regarded as possible, and the other should be for the more probable, safe amount. This will provide the basis for two estimates. These two figures can then be considered in relation to the market conditions, and the amount of business risk can be recognized, and evaluated.

WARNING! *It is seldom advisable to use an hour-per-standard-hour factor that is lower than the plant or department average.* There are, however, elements that may warrant such low factors, for example, larger than average quantities and more efficient tools.

IMPLEMENTATION ESTIMATING

Estimating implementation is also a part of the integrated estimating system. The purpose of this method is to assist the estimator in handling the main portion of a program. It is not to be used like a machine to mechanically produce an estimate, but is for use as a means of maintaining consistency.

It must be understood that this method provides the average, full measure of implementation, and allows for the use of a normal amount of standard implementation. If major items of implementation from previous projects are to be used, the value of such items must be deducted. Conversely, if the program requires items of an unusual nature, the value of such must be added. Specifications must be considered, and if there are unique requirements, the totals from the charts (Figs. A3-41 to A3-49) must be adjusted accordingly.

For other than a fully planned and estimated program, this method provides a reliable approach to estimating when used for the particular field and firm that provided the basic data for its development. It is also advisable to use this method to check an implementation estimate on a fully-planned program.

There are definite reasons why this method is good. It is based on the standard hours as developed for the product by the use of the charts (Figs. A3-5 to A3-27) in the product division of method 5, which provides a consistent measure of complexity for the various types of work. With a consistent complexity rating available, it is possible to evaluate a new project in respect to costs of previous programs.

The historical elements from previous programs are built into the implementation charts (Figs. A3-41 to A3-49) by plotting typical estimates and checking them by actual implementation costs, and by parallel use with estimates made by detailed planning and costing.

CHART IA.   IMPLEMENTATION ESTIMATING
100/MONTH RATE (LOW RANGE)

TOOL DESIGN

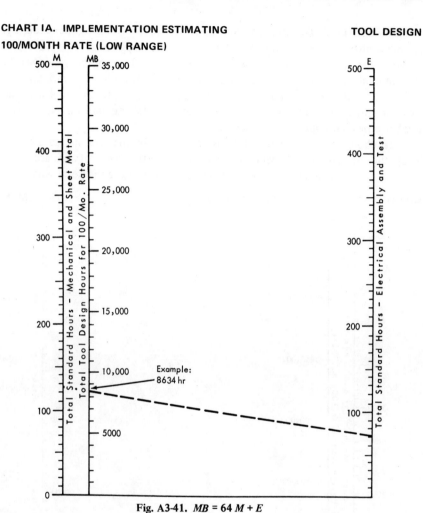

Fig. A3-41. $MB = 64\,M + E$

Basic Outline

Implementation requirements have been divided into two groups:

1. Tooling.
2. Test equipment and planning.

Tooling is more closely related to the mechanical and sheet metal standard hours than to the electrical and test standard hours. The reverse is true for item 2, test equipment and planning.

Method of Use: Tooling. The tool design chart for 100 items per month is provided in two parts. Figure A3-41 is for the low range of standard hours as required for small programs, and Fig. A3-42 is for large programs. (NOTE: Roman numerals are used for chart numbers in implementation estimating.)

The first step is to summarize the standard hours for the product into the two categories: (*M*) total mechanical and sheet metal standard hours, and (*E*) total electrical assembly and test standard hours.

Then choose Fig. A3-41 or A3-42 as determined by the total standard hours; locate the mechanical standard hours on scale *M* and the electrical standard hours on scale *E*. A straightedge through these points will locate the point where the

CHART IB. IMPLEMENTATION ESTIMATING   **TOOL DESIGN**
100/MONTH RATE (HIGH RANGE)

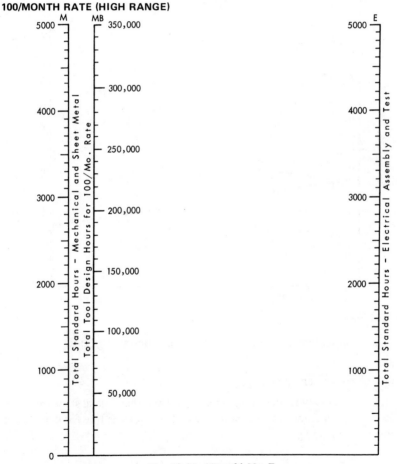

Fig. A3-42. $MB = 64 M + E$

tool design hours for a rate of 100 per month is read on scale *MB. Note this number; it is the base figure used in establishing all the tooling costs.*

The tool design hours *for the program* are determined by the use of Fig. A3-43, as follows. The basic tool design figure for 100 items per month as determined by Figs. A3-41 and A3-42, and the program rate of production per month, are located on scales *MB* and *R,* respectively, on Fig. A3-43. A straightedge through these two points determines the amount of tool design hours for the program on scale *X,* Fig. A3-43.

CHART II. IMPLEMENTATION ESTIMATING

PROGRAM PRODUCTION RATE

TOOL DESIGN

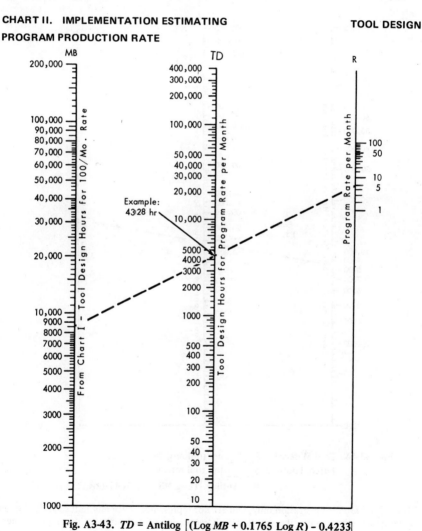

Fig. A3-43. $TD = \text{Antilog}\left[(\text{Log } MB + 0.1765 \text{ Log } R) - 0.4233\right]$

Tool manufacturing hours for the program are determined in the same manner, using Fig. A3-44.

Total tooling material is determined by multiplying the number of tool manufacturing hours by $3.00* per hour. Purchased tools are equal to the number of tooling hours times $6.00* per hour.

CHART 111. IMPLEMENTATION ESTIMATING　　　　**TOOL MANUFACTURING**
PROGRAM PRODUCTION RATE

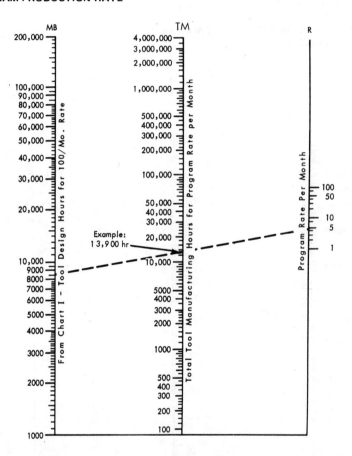

Fig. A3-44. Tool Material @ $_____ per tool mfg. hr.
　　　　　　Purch. Tools @ $_____ per tool mfg. hr.

$$TM = \text{Antilog}\left[(\text{Log } MB + 0.17647 \text{ Log } R) + 0.0835\right]$$

---

*These rates are for illustration only, and are not representative for any industry. Presented here as a method, only, the charts can, however, be of assistance in developing a new approach to estimating implementation if they are adapted to an individual firm's needs and experience records.

Test Equipment and Planning. Test engineering for a production rate of 100 per month is developed by the use of Figs. A3-45 and A3-46 in the same manner as was used with Figs. A3-41 and A3-42 for determining *MB*. *Note this figure (EB); it is the base figure used in determining the test equipment and planning hours.*

This basic figure (*EB*), and the program rate, can be used with Fig. A3-47 to determine the test engineering hours for the program.

CHART IV. IMPLEMENTATION ESTIMATING   **TEST ENGINEERING**
100/MONTH RATE (LOW RANGE)

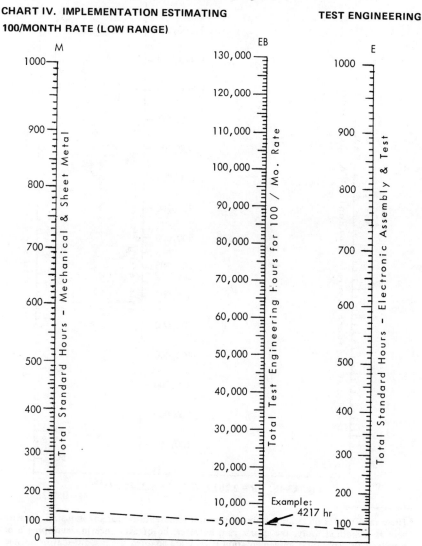

Fig. A3-45. $EB = 2.1619 \, M^{1.435} + 4.2344 \, E^{1.435}$

Figure A3-48, the basic figure (*EB*), and the program production rate determine the test equipment manufacturing hours for the program.

The test equipment material is determined by multiplying the test equipment manufacturing hours by $45.00* per hour.

CHART IVB. IMPLEMENTATION ESTIMATING　　　　　　　**TEST ENGINEERING**
100/MONTH RATE (HIGH RANGE)　　　　　　　　　　EB

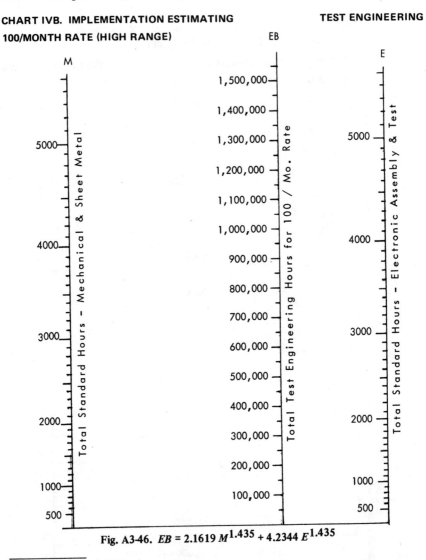

Fig. A3-46. $EB = 2.1619\ M^{1.435} + 4.2344\ E^{1.435}$

---

*These rates are for illustration only, and are not representative for any industry. Presented here as a method, only, the charts can, however, be of assistance in developing a new approach to estimating implementation if they are adapted to an individual firm's needs and experience records.

Figure A3-49, the basic figure *(EB)*, and the program production rate determine the planning hours for the program.

An example of implementation based on these charts is given in Fig. A3-50. This is an estimate of the required implementation for the black box used in the previously presented example.

CHART V. IMPLEMENTATION ESTIMATING　　　　　　　　**TEST ENGINEERING**
PROGRAM PRODUCTION RATE

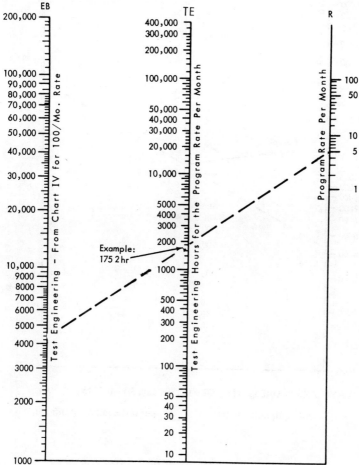

Fig. A3-47. *TE* = Antilog $\left[(\text{Log } EG + 0.285 \text{ Log } R) - 0.582\right]$

**CHART VI. IMPLEMENTATION ESTIMATING
PROGRAM PRODUCTION RATE**

**TEST EQUIPMENT
MANUFACTURING**

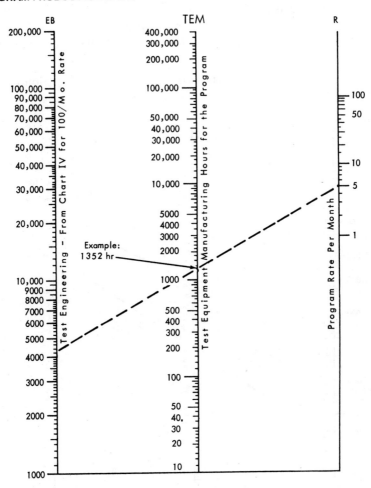

Fig. A3-48. $TEM = \text{Antilog} \left[(\text{Log } EB + 0.3584 \text{ Log } R) - 0.7445 \right]$

Test equipment material @ $_____ per test equipment mfg. hr.

CHART VII. IMPLEMENTATION ESTIMATING   **PLANNING**
PROGRAM PRODUCTION RATE

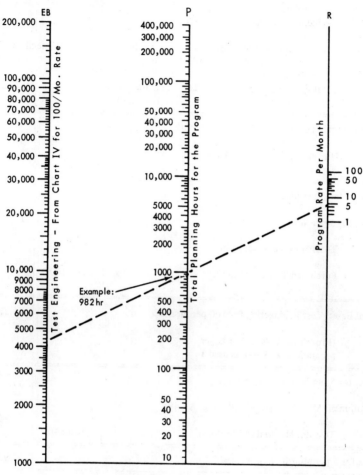

Fig. A3-49. $P = \text{Antilog} \left[(\text{Log } EB + 0.1263 \text{ Log } R) - 0.721 \right]$

IMPLEMENTATION ESTIMATE		
Mechanical Standard Hours (See Fig. A3-3)		128.2 (M)
(Composed of fabrication and assembly hours for both mechanical and sheet metal plus the fabrication for the electrical.)		
Total Electrical Assembly and Test Standard Hours (See Fig. A3-3)		71.5 (E)
Mechanical Base: Tool Design for 100 per Month		8,634 hr
(From Fig. A3-41. For larger programs Fig. A3-42 is used.)		
Tool Design for 5 per Month–Fig. A3-43.		4,328 hr
Tool Mfg. for 5 per Month–Fig. 3-44.		13,900 hr
Total Material @ $3.00/hr	$41,700	
Purchased Tools @ $6.00/hr	$83,400	
Electrical Base: Test Engineering for 100 per month		4,217 hr
(From Fig. A3-45. For larger programs Fig. A3-46 is used.)		
Test Eng. for 5 per Month–Fig. A3-47.		1,752 hr
Test Equipment Mfg. for 5 per Month–Fig. A3-48.		1,352 hr
Test Eq. Material @ $45.00/hr	$60,840	
Planning for 5 per Month–Fig. A3-49.		982 hr

Fig. A3-50. Typical Method V Implementation Estimate–for the Black Box.

Index